IEE Management of Technology Series 7

Series Editor: G. A. Montgomerie

PERSPECTIVES ON PROJECT MANAGEMENT

LPL

Mr R. N. G. Burbridge, OBE

Ron Burbridge has worked in the electricity supply industry for England and Wales since 1948. He has been involved in engineering design, project management and commercial work on both conventional and nuclear plant. As a Project Manager for the Central Electricity Generating Board he constructed the oil-fired power stations at Pembroke and the Isle of Grain. As Director of Projects he was responsible for the CEGB's Dinorwig, Drax and Littlebrook power station projects. He is currently Divisional Director of the CEGB's Generation, Development and Construction Division and Deputy Chairman and Chief Executive of the Project Management Board for the Sizewell 'B' PWR nuclear power station.

He has served as a Member of Council, Chairman of the Management and Design Division, and a Member of the Science, Education and Technology Division of the Institution of Electrical Engineers. He is a past member of the National Economic Development Committee for Engineering Construction. He is currently a Member of the Project Management Forum and a Director of the Major Projects Association, Templeton College, Oxford.

This volume includes a comprehensive introduction to the subject written by the editor.

PERSPECTIVES ON PROJECT MANAGEMENT

Edited by

RNG Burbridge

Peter Peregrinus Ltd on behalf of the Institution of Electrical Engineers

Other volumes in this series

Volume 1   Technologies and markets
           J. J. Verschuur
Volume 2   The business of electronic product development
           Fabian Monds
Volume 3   The marketing of technology
           C. G. Ryan
Volume 4   Marketing for engineers
           J. S. Bayliss
Volume 5   Microcomputers and marketing decisions
           L. A. Williams
Volume 6   Management for engineers
           D. L. Johnston

Published by: Peter Peregrinus Ltd., London, United Kingdom

© 1988: Peter Peregrinus Ltd.

British Library Cataloguing in Publication Data

Perspectives on project management.—
(IEE management of technology series; 7).
1. Industrial project management
I. Burbridge, R.N.G. II. Series
658.4'04 HD69.P75

ISBN 0 86341 105 3

Printed in England by Short Run Press Ltd., Exeter

Contents

Summary

Introduction: Historical and Contemporary Perspectives

R. N. G. Burbridge,
Divisional Director,
Central Electricity
Generating Board,
Generation Development and
Construction Division

1 The Compleat Project: the Successful Pursuit and Development of Projects

D. M. Slavich,
Vice President and
Chief Financial Officer,
Bechtel

The author reviews the changing international sources for project finance and analyses the circumstances in which capital investment in a successful project is likely to occur. The planning and regulatory aspects of developing a project, together with its financing, risk analysis and contractual strategy are briefly considered. The growing world economic inter-dependence and consequent inter-nationalisation of many projects is emphasised.

2 Project Management — A Client's View

Sir Alistair Frame,
Chief Executive,
Rio Tinto Zinc

The author outlines the approach adopted by the RTZ Group in evaluating and managing large construction projects throughout the world. Methods of assessing new capital investments and the benefits of post-project audits are explained. The procedures adopted by the client in developing a project as well as in its subsequent execution are summarised. The importance of selecting the right Engineer or Managing Contractor with appropriate procedural arrangements is emphasised, as is the need to appoint the right Project Manager. Illustrative examples are quoted of the problems that can occur when these strategies are not followed and of the difficulties in identifying those rare individuals with the personal capabilities that are demanded by successful project management.

3 The Project and the Community   S. C. Goddard,
Corporate Director of
System Planning,
Central Electricity
Generating Board

The author analyses the impact of large scale construction work on the local community and the environment with reference to the power station construction projects of the Central Electricity Generating Board and its "Good Neighbour Policy". The process leading to the granting of consent for new stations in England and Wales is described and the importance of consultation with the local interests involved, together with the subject matter of such consultation, is reviewed.

The latter part of the chapter explores those matters associated with a major project that may be considered by the local community to be of benefit, as well as those where disadvantage is perceived to occur. A degree of local acceptance and support and the need to establish good community relations are seen as significant objectives for the developer and his project management.

**4 Some Thoughts on High Budget
Projects**   A. J. O'Connor,
President,
New Brunswick
Electric Power Commission

The author considers the uncertainties that surround the completion of major projects in North America and analyses the problems currently associated with civil nuclear construction. The significance of the owner/client role in successful project management and the development of proper organisational structures is noted and the type and form of contract most suited to nuclear construction in North America is reviewed.

The author argues that firm price contracts are only appropriate if the design is complete at time of tender; as this is rarely possible for nuclear construction reimbursable forms of contract are preferable as is the employment of a managing contractor as construction manager. Whatever form of contract is employed the development of trust and confidence between owner and contractor is considered to be essential, as is the establishment of good labour relations through the use of project agreements. The need for projects to win acceptance locally and to provide economic benefit without sacrificing safety standards or the environment is seen as the most important future challenge.

5 The Management of Joint
   Venture Projects

C. Fleming,
Manager,
Construction Division,
BP International

The author describes the nature of joint venture projects and why the requirements of managing them are different to those applying for sole owner projects.
The reasons for the increasing trend towards joint ventures for large capital
projects are discussed together with the effects arising from the strains and
tensions evident when a group of companies of varying sizes and financial
strengths, each with differing priorities, reasons and interests, agree to par-
ticipate in the execution of a major project.

Various options for managing such projects are given and many of the
difficulties encountered by the project management team are considered. The
importance of establishing a relationship of mutual trust and respect between
the project management team and all the participants is stressed as a key factor
in achieving successful completion of the project.

6 Contract Strategy

F. Griffiths,
Director,
Frank Griffiths Associates

The author reviews all the main factors that require consideration in developing
a contract strategy relevant to the particular requirements of an individual
project. The role of estimating in the contract process and risk sharing are
considered and examples of certain less than successful projects are raised. The
major portion of the chapter provides a comprehensive survey of those matters
influencing the type and conditions of contract that may be chosen for a project
as seen from the client's view-point.

7 Turnkey versus Multi-Contract

D. C. E. Brewerton,
Manager Project Services,
Kennedy and Donkin

The author reviews the importance and objectives of the purchaser as seen
through the eyes of a consulting engineer with experience in managing projects
on a turnkey basis. An example is given of how both turnkey and multi-
contracting operates and the paper reviews many of the advantages and disadvantages inherent in each system. The implications in adopting either system
at the tender stage are also raised together with observations on executing either
form of contract on issues such as the duration of programme, collection of

information, access, monitoring site control and cost implications, as well as quality and technical performance. The duties and responsibilities of both purchaser and consultant are also reviewed and the relevant merits of each system are summarised in terms of time, cost and quality.

8 Quality Assurance and **Project Management**	**R. M. Macmillan,** **Policy Development Manager,** **Central Electricity** **Generating Board**

The author describes the CEGB Quality Assurance policy and the practices adopted corporately and by individual responsible officers to ensure that the required quality of plant and equipment is correctly specified and obtained. The chapter reviews experiences on new power station projects and also describes developments taking place for the assurance of quality during plant operation.

9 Quality Assurance	**C. A. Mills,** **Westinghouse Canada Inc** **(Retired)**

The author describes the use of the Project Quality Plan as a means of defining specific quality programmes for major military and high technology commercial and industrial contracts. The value of using a Project Quality Plan is outlined, together with some comments on the timing of its preparation, methods of monitoring the performance of the plan are described.

10 Computer Applications	**D. Langan,** **Director, Taylor Woodrow**

The author outlines the physical characteristics of North Sea oil projects and their effect on the design, implementation and improvement of computer-based information systems. The development history of such systems is described as is their scope and content, together with an explanation of the inter-relationships of those systems dealing with project planning, materials procurement and financial control. Attention is drawn to the flexibility of the system in responding to the needs of client organisations with their different managerial styles. The author also draws upon his early experiences in developing the system. He emphasises the importance of user training and user involvement in system design to ensure the eventual acceptability and success of the system. The ability of the system to accommodate a situation in which project design alters during the construction stage is also noted.

**11 Essentials in
Project Management**

**H. Masding,
Retired Policy Development
Manager, Central Electricity
Generating Board**

The author sets out the principal objectives in the management of industrial projects and the example chosen is the design and construction of power stations. Some consideration is given to the need for the replacement of existing generating plant in order to demonstrate the possible future work load for both the manufacturing industry and utility. The importance of the contract strategy is emphasised and the options open to the purchaser are considered in general terms. The paper also gives an outline of the functions of the purchaser, the engineer, the design and build contractor and the project manager. The paper disputes the viability of some of the current conceptions, the main theme however, is to stress the importance of completion to time and cost and gives a view on how this is most likely to be achieved.

**12 Project Management —
A Review**

**J. W. Currie, Chief Engineer,
Generation, Design and
Construction
R. M. Gove, Chief Engineer,
Transmission/Distribution
A. F. Pexton,
Director of Engineering
South of Scotland
Electricity Board**

The authors provide a descriptive analysis of the strategies and procedures adopted by the South of Scotland Electricity Board for the management of large power plant projects. The use of consulting engineers and the development of the necessary skills and resources within the client organisation are described in the context of the restructuring of the British nuclear industry during the last 20 years. The changes in contract strategy which such restructuring has occasioned are also reviewed, as are the tender assessment procedures applied to the letting of negotiated contracts.

The highest priority is accredited to relations with Local Authorities and amenity bodies. The growing importance of the planning approach and consent process to project management is considered and it is suggested that the engineering institutions may be able to play a part in this process by assessing technical issues of national importance.

The SSEB's project management policy is outlined in the second part of the chapter. The importance of design contracts, replication, off-site assembly, key date reviews, site industrial relations and local welfare considerations are emphasised; as is the need for prompt and effective feed-back to secure improved plant performance and availability.

PM is + client orientated now
↳ "impatient clients"

Historical and contemporary perspectives: An introduction

Editor: R. N. G. Burbridge OBE
Divisional Director, Central Electricity Generating Board

Project management like politics is very much 'the art of the possible'. An appreciation of the skills required and the challenges faced in the construction of a major project is best achieved through the views and opinions of recent practitioners in the field. As an officer of a major British client organisation, it will be no surprise to the reader that this book is oriented towards a client's perspective on the subject, although the contractor's viewpoint has not been excluded. The latest matters of moment in the subject have been given prominence in the Chapters that follow, including the attention that is now given to the environmental impact of construction, the importance of quality assurance and the application of computer techniques. The perennial debates as to the most suitable forms of commercial relations between client and contractor and the most appropriate organisation for a project have not been omitted.

Many of these issues are not new. Indeed, it would be wrong to neglect the historical antecedents of the practices and procedures we now call project management. They date in part from antiquity as well as from the more recent developments in technology and organisation which we know as the Industrial Revolution. Such historical observation provides an added perspective to a study of contemporary project management, and will serve as an introduction to some fundamentals of project management considered in this book.

Origins of project activity

In introducing the separate topics contained in later Chapters, it is fitting that emphasis is given to one universal tenet of project management, namely the importance of the human factor. The hallmark of project engineering in every age from the ancient engineers to modern constructors is the requirement to organise and co-ordinate group activity, often involving very large numbers of people in widely scattered locations who may have different and conflicting objectives. This characteristic holds true regardless of the nature of the project, the age in which it is constructed or the political, social and economic system which generated the impetus for the project works. The management of this activity demands personal leadership of the highest calibre, for project management success can only be achieved through organising the efforts of others. It is not too much to claim that one hallmark of civilisation is the ability to engage in group activities for the execution of major projects, be they tombs and temples or manned flights into space.

Reasons for initiating large scale 'mega projects' have varied considerably down the ages, reflecting the moral and ethical values of a society, its political institutions, social organisation, wealth, technical development and many other factors, including matters of prestige and international relations. The impact of religious beliefs, for example, upon project works and their management should not be underestimated. Considerable resources were devoted by the ancient Egyptians and Greeks to their pyramids and temples, and probably the first engineer and architect known to us by name was Imhotep who built the first pyramid. Economic considerations, especially the securing of food supplies and the means of rapidly transporting such supplies and trade goods, have always been a significant project sponsor from ancient to modern times. The first Suez Ship Canal connecting the Mediterranean and Red Seas was started in the 6th century BC and ran east–west from a branch of the Nile before turning north–south around the Bitter Lakes to the Red Sea. It was eventually completed under Darius the First and lasted intermittently until the 8th century AD when it fell into disrepair.[1] Communications between the two Seas were not re-established until completion of the Suez Canal in 1869.

Equal in importance as an impetus to large-scale construction are matters of defence, international prestige and rivalry between nation states. The Great Wall of China, Offa's Dyke and the Maginot Line, to name but a few, all represent considerable feats of engineering and organisation. The impetus given to the American programme of space exploration by the initial Russian successes is obvious. It resulted in the expenditure of considerable financial resources and elite engineering manpower in this endeavour by both nations. The Apollo programme has been estimated to have cost $47 billion and absorbed 100,000 man years for each year of its duration. Subjecting such expenditure to the rules of business economics would clearly be inappropriate. The advancement of science and the future benefit of mankind are always difficult entries to evaluate on a balance sheet, as is the political credibility of one nation state compared to another. Such mega projects are clearly undertaken for socio-political reasons and consequently they may have a very broad impact upon a society and its neighbours in a psychological and even cultural sense.

A very different example from the space programme was the construction a century ago of the American Statue of Liberty in New York harbour. The statue was a gift by the French to commemorate the bond between the two nations, to celebrate the freedoms enshrined in the American constitution and France's role in assisting the thirteen colonies gain their independence a century earlier.

However, there was undoubtedly a further reason; by extolling the virtues of equality, liberty and political freedom in the American constitution, the original benefactors were advertising the lack of such freedoms in contemporary 19th century France. A hundred years later, project management activities are still endowed with political purpose within Europe. Joint ventures between member states of the European Community, particularly in the aerospace field, are regarded as ways of integrating the Community's manufacturing and commercial activities to produce sufficiently large and effective organisations able to compete internationally with concerns in North America, Japan and elsewhere. The leadership demands of a large multi-lingual, multi-national project organisation involving a plethora of different activities are perhaps greater than in any other identifiable managerial role. The way in which the organisation is structured to encouraged such effective leadership is considered in several later Chapters, and particularly in that entitled 'The management of joint venture projects'.

Such political motivators can just as easily inhibit as promote project work. The engineering capability to construct a tunnel under the English Channel has been available for decades, but no tunnel has been built despite the obvious advantages of a fast all-weather road and/or rail link to France. Hopefully, this project may now be completed in the next few years. British membership of the European Community and the dominance of our trade links with the Continent have undoubtedly played a part in promoting this development. The flood of seasick holiday makers across the Channel each year may have been a weightier consideration in the minds of promoters and government than the time taken for our exports and imports to complete their journeys. The involvement of the state in the decision-making process concerning large projects is central and the political motivation of the politicians in power is crucial, but so too are the sentiments of the general public who elect such officers to power.

The role of the media in forming public opinion on such matters must not be under-estimated. Project managers must come to terms with 'trial by television' when trades unions, politicians, environmentalists, TV commentators, and others query our policies and procedures in the public eye. The widespread publicity accorded all matters associated with the protection of the environment is recognised and catered for in the United Kingdom through our planning statutes and regulations, public inquiry procedures as well as through Parliamentary debate. The heightened public awareness of the vulnerability of the environment to industrial processes of all descriptions is raising many issues. Every stage in our planning, design, construction and operation of industrial plant is likely to come under close and constant scrutiny as the civil engineering industry knows only too well. The longest ever Public Inquiry into the proposed Sizewell B PWR nuclear power station took two years to complete, and the report of the Inspector who ran the Inquiry took over a year to present. Public anxiety demands that client operators and construction companies actively plan to alleviate such concern. An explanation of the way in which my own organisation approaches this task is contained in the Chapter entitled 'The project and the community'. Good community relations must be accorded the highest priority by the project manager, but so too must the general need to inform and advise the public at large of the rationale behind a project, its benefits, the steps taken to minimise disturbance and disruption during construction and operation, and the safety and security of that operation.

In emerging countries in the Third World these issues are exacerbated by the desperate shortage of well directed finance for development purposes. The very large increases in energy costs during the last decade, and the resultant recession in many industrialised economies, severely aggravated the endemic economic problems of the Third World. The loss of oil revenues and the fall in commodity prices in this decade have again engendered financial difficulties for developing countries. These pressures have led the governments of certain nations to seek 'build, own and operate' bids from contracting organisations. The increased cost and financial risk associated with these arrangements and mega projects in general has led to the greater frequency of joint ventures as a means of spreading that risk. Growing world economic interdependence and the internationalisation of project financing and execution are developed in the first Chapter following this introduction.

① slave / master

Management and men

If the hallmark of project management is the need to organise and co-ordinate group activity, the means whereby the individual and the group are managed and motivated are of critical importance. The use of a stick to wallop the laggardly has been a traditional part of managing a gang of Egyptian labourers from ancient to almost modern times. The architect Nekhebu, in boasting on his tomb of his many virtues and kindnesses, mentions the fact that he never struck a workman hard enough to knock him down.[1] In China the main work of linking the parts of the 1200 mile long Grand Canal was done around 600 AD. It is reported that 5.4 million labourers were conscripted for the work guarded by 50,000 policemen, a ratio of one policeman to just over 100 labourers. Shirkers were beheaded when caught and two million died during construction.

In the early days of the Industrial Revolution motivation and discipline depended upon traditions of craftsmanship and the master/servant relationship.[2] This proved inadequate for the demands of the new industrial enterprises, and various combinations of reward and punishment were adopted to suit the requirements of the industrial process or the dictates of the proprietors or their agents. Not surprisingly, relations between masters and men were diverse. For construction work labour was supplied by a ganger to a subcontractor to carry out a defined piece of work. The system can be traced from the canal-building era in England and continued into the period of railway construction, and continues today in the form of 'labour only' subcontracting. Payment was often made through the 'butty' system, whereby each gang struck a bargain with a subcontractor to complete a piece of work, then generally shared the pay equally with each other. Alternatively, a fixed daily rate may have been paid directly to each man by the subcontractor, or piecework may have been made available (so much pay per foot excavated).[3]

The discipline, or rather the ill discipline, of the railway navvies was legendary even in their own time. Yet the successful contractors of the period, notably Brassey and Peto, could inspire great loyalty and respect in this toughest breed of men. Peto was unusual in that he employed the men directly, paid them weekly and was deeply concerned for their sobriety and safety. Brassey believed it best to give the men a personal (financial) interest in completing the contract, and preferred the butty system. He paid the highest wages, did not exploit the labour, was never unnecessarily cruel or even hard. If the men were ill or suffered an accident he supported them; if they died he helped their dependants; and when times were bad he tried to spread his contracts to minimise the effects of unemployment.

The construction of canals, railways, docks and certain military structures probably represented the largest concentrations of capital during the eighteenth and nineteenth centuries, and the civil-engineering contracting industry slowly developed to meet the need. The achievement of men like Brassey and Peto was the creation of a new type of organisation. Previously, small craftsman firms predominated. For example, construction of the Middlesex Canal in late 17th century North America was undertaken mainly by the yeomanry, small landowners and farmers, who contracted to build specific lengths of the canal.[4] While experienced in digging ditches and draining meadows, such men had no significant financial resources or appreciation of the needs of working to programme. In the building industry in England, the traditional method of contracting was for one craftsman to accept responsibility for a job, sub-contracting any specialist activities to others, and for these roles to be frequently reversed.

In contrast, Brassey gradually assembled a huge and complex organisation with a vast labour force of up to a hundred thousand men, scattered over half the world's continents.[5] The organisation was pyramidal in nature with Brassey himself alone at the top. Next were his agents, a carefully selected group of highly trusted men who in modern parlance were the project managers of the organisation. These agents managed the next strata — the subcontractors who were responsible for a part of the contract or project — while under the subcontractors were the gangers, each in charge of a gang of about a dozen men. There were many reasons for Brassey's success, but one of the key ingredients was his careful selection of a superb team of capable and reliable agents. Several Chapters in this book emphasise the importance of selecting and developing that key group of project managers and of providing the right organisation through which their skills may be exercised.

Culture and productivity

The factors that characterise modern project management, as opposed to the efforts of our forefathers, are the requirements to construct to time and cost. Growing technical complexity of the equipment we construct has not rendered this process any easier, and the erection of plant on the boundaries of engineering science creates a high degree of financial and technical risk. The importance of completing projects to time and cost is axiomatic. If a trading economy cannot renew and update its means of production in as cost effective a manner as its competitors, then it must inevitably lose market share, and thereafter the ability to compete at all will be steadily eroded. The importance of good labour relations and the achievement of high levels of productivity are as important in securing a cost-effective performance as the establishment of a secure design in advance of manufacture, the constructability of the plant, the choice of appropriate commerical arrangements, a relevant organisation and personnel and all the other ingredients for a successful project.

Labour relations are of course highly variable and, in the past, construction and other industries have not always been noted for their harmonious relationships. The penalties of poor working relations can be harsh in the extreme, as the President of the Institution of Mechanical Engineers has stated:

> Trades once monopolised by England, the whilome workshop of the World, have become wholly or partially settled elsewhere; whilst our own working people, still unshaken in their belief in the virtue of strikes, high wages, short hours, and workshop restrictions, are only partially employed . . . till in the end . . . we shall have lost the large part of our foreign export trade . . .[6]

These remarks were made in 1877. A century later Britain has certainly seen the demise of many manufacturing industries, from motor cycles to space vehicles, which are now 'settled elsewhere'.

However, the UK in the present decade is generally enjoying a period of excellent labour relations, both on our construction sites and in manufacturing industry. On British mega projects this is due in no small measure to the active assistance and co-operation of the trades unions concerned, which is particularly noteworthy during a time of growing unemployment and falling union membership. Recent legislation has assisted this process by supporting trades unions in conducting their own affairs in a proper manner. Legal requirements for unions to ballot their members before striking, or before closed shops can be established, and other measures have so altered the climate that full-time union officers can now exercise leadership in a responsible and effective manner to the benefit of their members and the industries that employ them. We cannot conclude, however, that labour militancy has now disappeared and that it cannot return if different conditions pertain in future. The need for effective project management and sound labour relations remains a prime requirement for undertaking project work.

Current economic circumstances in Britain have also ameliorated the problems facing project management by both clients and contractors. The drop in the rate of inflation has allowed costs to be more closely controlled, reduced an area of substantial risk and encouraged clients to seek firmer priced contracts from their contractors. The diminution in the number of major investment projects undertaken at any one time has allowed a closer control over the industry as a whole compared to the seventies, when an 'over heating' of the construction industry occurred as numerous projects, especially in the energy sector, were undertaken at the same time. High levels of unemployment have eradicated competition for labour, and skill shortages are generally no longer a significant restriction. The nature of site work also appears to be changing: the greater use on site of high-cost mechanical equipment, and the wider adoption of off-site prefabrication, are both reducing the work content of site erection and altering the nature and quality of the labour skills required.

The ability of British project managers to complete programmes to time and cost is analysed in the last two Chapters. Considerable success has been recorded this decade in the construction of major new power stations by the electricity utilities. The Chapters provide a background to contracting work in the electricity supply industry together with descriptive analyses of the strategies and procedures currently adopted in project managing such major works. But we are not and must not be complacent. Power plant construction performance in the Far East, notably in Japan and Hong Kong, still exceeds European achievements. The reasons for this Japanese performance record are many and complex, and may have as much to do with Japanese culture and tradition as managerial practices and commercial organisation. Any analysis is therefore likely to conclude that the Japanese are exellent constructors because they are Japanese, and that any attempt to apply Japanese methods in a British or European context may even be harmful on the grounds that what works for the Japanese may prove disastrous for the European. There are, however, certain aspects of the construction industry in Japan that do warrant attention.

Firstly, it is doubtful whether the Japanese construction worker actually works harder or faster at a personal level than our own. What he does do is to work productively for longer periods of time. The normal working week is 48 hours and 60 hours is not unusual in a standard 6 day week. British experience suggests the adoption of a 50- or 60-hour week for three or four years does result in much unproductive time being paid. Indeed, with rising unemployment, union opposition to extensive overtime may increase. The answer must be the introduction of shift working that provides a working week of around 40 hours for the individual, but allows the liquidation of anything up to 168 programme hours per week using 2-, 3- or 4-shift teams.

The largest opportunity to improve our productivity lies in the improvement of work design. It means paying far greater attention at the design stage to the constructability of our plant, as well as to the adoption of improved erection methods and equipment. Such items as the greater use of mobile scaffolding, larger cranage, coupled with a higher proportion of prefabricated plant modules, holds out the promise of much higher labour productivity and cost-effective performance.

Other aspects of the Japanese industry are also noteworthy. The client awards his contract to a manufacturing company, which is usually a very large engineering corporation, employing a great range of technical expertise within its organisation. That company usually manufactures the whole of the plant and accepts a turnkey-type arrangement for the undertaking. Such contracts are firm price and Japanese contractors seem to regard the submission of a claim as a 'loss of face'. Japanese corporations can undertake enterprises on this basis because of the stability of the Japanese economy and their size. The manufacturer in turn subcontracts the erection work, but not in the sense of relinquishing authority for that work. He continues to provide the majority of the site technical and managerial personnel down to and including the labour foremen. The erection subcontractor provides the labour, and no doubt managerial expertise and technical knowledge on erection methods.

The erection subcontractors are also very different. They are major corporations in their own right and they supply *all* trades skills required at site, who are engaged on the same terms and conditions of employment. Such conditions do not include the provision of incentive bonus schemes on site. The only form of bonus is an annual profit-sharing arrangement, which is not subject to negotiations between company and union but is determined solely by the company and related to its overall financial performance. Japanese union negotiations (where they exist) relate to base rate, but, more importantly, their unions see their role as being supportive of the company's endeavours and adversarial relationships are not prevalent. This erection subcontract system results in up to 70% of a power-station project's work force being permanent employees of their companies, and in Japanese terms that means a near guarantee of lifetime employment. Construction work also appears to enjoy high social status, and this factor, coupled to Japanese veneration of their employer, the all pervading work ethic and a very competitive work environment, ensures that considerable talent and expertise are devoted to achieving success on site as elsewhere.

These attitudes are very different to those that sometimes prevail in the older industrial nations, some of which may be very deeply rooted. For instance, the I. Mech. E.'s President quoted above, speaking in 1877, stated that the working man had:

> . . . become thoroughly imbued with three transparently false notions. *First*: That he is entitled to share in his employer's success in business.

Second: That he has a right, in combination with others, to exact conditions from and prescribe terms to his employer.

Third: That he has a right, in combination with others, to force his employer to yield to his exactions and prescriptions . . . which are so exercised as to control the employer in regard to the manner in which he shall conduct his business, and even in respect of his relations to the general public at home and abroad.[6]

Seen from this perspective our attitudes have undoubtedly changed. Many employers actively seek arrangements whereby their employees share in the company's success and few deny that unions have a role in representing their members. Enlightened unions do not seek to circumvent a management's right to manage, but the creation of improved institutions and procedures for managing labour affairs remains a challenge. There is much evidence that this is being achieved. In Britain the creation in 1981 of a new National Agreement for the Engineering Construction Industry has provided the framework and impetus for higher productivity. The growth of single-union/single-status 'Japanese' style agreements in manufacturing industry demonstrates the emergence of highly co-operative attitudes.

Techniques and procedures

This volume is not a textbook on project-management techniques and disciplines. Excellent textbooks on this subject are available, including a recent publication by my own organisation.[7] However, this volume would have been incomplete without reference to the procedures and systems for measurement and control that are now advocated by the practitioners. In view of the increasing technical complexity of our projects, their increasing scale, and the prospect, however remote, of catastrophic failure and consequent damage to the environment and its inhabitants, the role of quality assurance has assumed greater significance. Its ultimate purpose is to ensure the plant or structure performs in accordance with its specification throughout its operating life. In meeting these objectives a monitoring system of checks and controls must be established that generate information vital to the project manager. Two complementary Chapters on this topic are given, supported by a dissertation on computer-based information and control systems developed by a contractor for project work in the British North Sea.

Mega projects require the availability of extremely large amounts of finance, whose procurement is increasingly an international or governmental affair. The capital once allocated is very much at risk, especially during the construction period, and the appropriate allocation of that risk through the contract between client body and contractor is an essential prerequisite to successful project completion. Two Chapters on contract strategy and turnkey versus multi-contracting provide a comprehensive survey of those matters influencing the

the type and conditions of contract and the varying responsibilities of the parties under different forms of contract.

There are a number of other requirements that can be identified and which must be observed if effective management of a project is to be achieved.[8] For power plant construction, it is particularly important to maximise the completion of design work before the start of manufacturing and construction. Preferably, design work contracts should be awarded in advance of any hardware manufacturing commitments and the start of civil engineering works on site. They must be sufficiently resourced so that they are completed before commitment to site. Thereafter a design freeze, or near freeze, should be adopted to ensure the optimum completion of the manufacturing contract and to maximise the security of subsequent plant deliveries to site. Secure deliveries will in turn ensure that site erection can be planned in an optimum manner for the project as a whole, thus avoiding access difficulties by a variety of contractors to specific work areas. It should be added that the construction of well proven, replicate plant greatly facilitates both the design, manufacture and erection processes, reduces the financial risks and assists in matching the size and scope of individual contracts to the capabilities of the contractor.

A thorough vendor assessment by the client of prospective contracting companies is also a prerequisite for project success, in that it provides insurance that the financial, technical and managerial resources of the company are equal to the task of discharging the contract. The maximum use of firm price contracts, supported where necessary by various key date incentives and review procedures, ensures that all parties are committed to achieving completion of the contract works in time and to budget.

It is, of course, essential that the client or his project management nominee provides strong leadership in the overall design and management of the interfaces between contractors. Equally strong leadership is needed to manage the interfaces between the project and all those external influences emanating from government, various interest groups and society at large. Full authority should be delegated to a single project manager in charge of a multi-disciplined team, and adequate resources must be obtained to meet the programme obligations of all parties. At site, co-ordination demands the creation of a single body or authority with power to determine and implement all necessary policies and procedures to execute the project, especially in the man management field.

The final Chapter in this book expand this broad sweep of the range and complexity of the project manager's task. He must not only be an able manager and a sound administrator, possessing a thorough understanding of the science and engineering technology relevant to his project, but he must also be an exemplary politician, social scientist and (above all) communicator. Such individuals are rare and hard to find. Like morals, project management skills can only be 'caught not taught' and they must be experienced to be understood, although education can undoubtedly help this understanding. This book contains the distillate experience of several eminent practitioners who have contributed their thoughts in the interests of better project management. My thanks are due to them for their contributions and the opportunity to learn their views and opinions.

References

1 DE CAMP, L. S.: The ancient engineers (Sovereign Press, London, 1963)
2 POLLARD, S.: 'The genesis of modern management' (Arnold, London, 1965)
3 COLEMAN, T.: 'The railway navvies' (Hutchinson, London, 1965)
4 ROBERTS, C.: 'The Middlesex Canal 1973–1860' (Harvard University Press, 1938)
5 WALKER, C.: 'Thomas Brassey: Railway builder' (Cox and Wyman, London, 1969)
6 HAWKSLEY, T.: 'Presidential Address', *Proc. I. Mech. E.* 1877 **28**, pp. 167–175
7 CEGB Generation, Development and Construction Division (Barnwood): 'Advances in power station construction' (Pergamon, Oxford, 1986)
8 BURBRIDGE, R. N. G.: 'Some art, some science and a lot of feedback', *IEE Proceedings*, 1984 **131**, Pt. A, pp. 24–37.

Chapter 1

The Compleat Project: the successful pursuit and development of projects

D. M. Slavich
Vice President, Bechtel Group Inc.

1.1 Introduction

The term 'compleat' is used, with due acknowledgement to Isaak Walton, because it connotes not only complete in the sense that there are no missing pieces, but also that which is balanced and well rounded with the parts fitting together in a manner which enhances its overall effectiveness.

Thus, the Compleat Project is one which includes not only the essential ingredients of a successful project, but one in which the parts are combined in a balanced and effective way with proper attention to the planning and execution of each of its major aspects.

It is understandable that each of us in viewing a project from the perspective of our own discipline tend to emphasize the importance of those aspects of the project with which we are most knowledgeable and familiar.

This tendency must be overcome if we are truly going to conceive, plan and implement the Compleat Project.

An important point which is often overlooked is that projects, like products, have to be marketed. They need to be made attractive to prospective lenders and equity participants, who have many alternatives from which to choose.

Moreover, the market for projects like the market for products is constantly changing. These changes, and the underlying factors influencing them, need to be monitored and understood.

1.2 Changing market for projects

The shifting forces impacting the international market for projects take many different forms.

The rapid accumulation of dollars and other hard currencies in OPEC countries in the 1970s reaching over $100 billion a year at the 1980 peak, provided the Middle East with the resources to fund a wide variety of projects in their countries ranging from schools to airports to the entire new industrial city of Jubail, and a multitude of other projects around the world as capital was transferred through global capital markets. Since 1980 a steady decline in oil prices has rapidly reduced these surpluses, with a consequent reduction in domestic spending in the Middle East and the near elimination of funding for projects in other countries from OPEC sources.

While this source of funds was dramatically declining, other sources were increasing as world economic activity ebbed and flowed. Most notably Japan's capital outflows, born of more than $30 billion a year trade surplus, have been a major factor in the rapid economic growth of the South-East Asian Region and the USA and have provided significant support for new projects throughout the world. Japanese capital investments around the world are now reported to be of the order of $50 billion per year.

In contrast, many countries are suffering from severe capital shortages. The mounting external debt problems of several developing countries in Latin America, Africa and South-East Asia, amounting to half a trillion dollars in 1984, has severely restricted the market for even badly needed projects in those countries. These countries had access to capital in the 1970s, but now are suffering from a severe shortage of funds. Markets for goods and services have also changed dramatically. In some cases this has resulted in the pursuit of non-traditional forms of project financing such as counter-trade and barter.

An ongoing assessment of the changing market for projects is critical for those who seek a role in their implementation. Equally important is an understanding of the essential characteristics and attributes of a Compleat Project, and how to nurture its development.

1.3 Features of a successful project

Here are the most important features of a Compleat Project:

First, the project must be needed. That is, it must serve a useful purpose which may be any of the following:

  (a) to provide vital infrastructure services required for further future development
  (b) to develop the nation's natural resources
  (c) to increase the worth of a commodity by manufacturing or processing it into a more valuable product or export or import substitution, or simply to exploit a profitable opportunity.

There are other reasons why projects are implemented such as enhancement of national prestige, but a project which is driven by economic needs and benefits is far more secure than one based largely on political support of the moment.

Secondly, the project should employ a commercially demonstrated technology which is in tune with the host country's stage of development and the expertise of the available labour force. The project should not be so technically sophisticated that there may be a shortage of trained personnel to operate it successfully.

Thirdly, the project should make basic economic sense. For example, if it is a manufacturing or processing facility, the project must be able to compete successfully on a worldwide basis over the long term. In the case of basic infrastructure projects such as power plants, a life-cycle economic evaluation must be made. Certain generating facilities such as nuclear plants may at first appear to be non-competitive owing to their high initial capital cost. However, when the low, stable, ongoing fuel and operating costs are taken into account, the apparent disadvantage often turns into a net long-term advantage.

Fourthly, the project sponsor/developer, whether foreign or domestic, must have the financial resources, credit-worthiness and staying power necessary for the undertaking.

Fifthly, other major project participants, such as the firm performing the feasibility analysis, the architect/engineer, the construction manager, the major contractors and suppliers and the equity participants and lenders, all need to be strong, reputable recognized firms in their particular fields prepared to make and stand behind appropriate assurances. This imparts strength and stability to the project, and bolsters its stature and credibility.

Sixthly, the host country must offer a hospitable climate for the undertaking, including a stable political environemnt, and demonstrate its ability and willingness to support the project's investors and lenders with assurances and guarantees, as necessary, without creating an excessive national debt burden. It is often helpful for the host country to be an equity participant in the venture.

These are the most important features of the Compleat Project. However, they do not carry the same relative importance in all projects. For example, in nuclear projects, compliance with host-country regulatory requirements carries a particularly heavy weight. Host-country support and participation play an especially important role in infrastructure and natural resource development projects.

1.4 Problems of project development

The process of developing the Compleat Project has become enormously complicated in recent years. Each major new project, it seems, generates coalitions of proponents and opponents. The proponents, led by the owners/investors, are often joined by labour, with no-growth/slow-growth advocates and environmentalists combining forces to speak for the opposition. In the case of any given project, several other concerned groups may enter the fray. The conflicts which arise between these opposing actions can pose formidable obstacles and cause frustrating delays. A particularly serious problem is often created by a myriad of overlapping government agencies that claim permitting or regulatory jurisdiction over a project, and seek to impose conflicting standards and requirements on it. If there are competing alternative projects, this, of course, further compounds the problem.

With this increasing array of problems confronting it, the Compleat Project must be well conceived and well planned in all of its aspects. This requires a diversity of disciplines, knowledge and expertise. It needs to proceed across a broad front, with the work of each member of an interdisciplinary team functioning concurrently and co-operatively with each other member to produce a co-ordinated, integrated whole.

Project development has become a time-consuming and complex process. New tools and imaginative approaches are needed to assure its success. Just as engineers use a critical-path method to control and monitor a project's design and construction, a similar integrated project plan is needed to monitor and control the total project development through its various phases, and keep the technical, social, governmental, institutional, economic, environmental and financial aspects of development coordinated and on track.

There is no single or easy prescription for piloting a project through the shoals of its development phases. The process will vary substantially from one project to another for a variety of reasons. However, it can be predicted with reasonable assurance that certain elements of the process are almost certain to require more time and effort than originally anticipated. These are:

Financing plant development: A comprehensive plan of financing the project needs to be developed. This is not a hypothetical plan but one expressly tailored to the project which can be implemented as the various aspects of the project are firmed up. It identifies the sources of equity and debt, the terms of the financing and the project cash flows needed to service this debt and provide a rate of return to potential investors. The ability to respond promptly to the need for financing a project requires a constant monitoring of worldwide commercial and export credit-financing sources, international financing agencies, financing mechanisms and the terms and conditions prevailing for particular countries and types of projects in the international sources of goods and services for the project and their associated terms and conditions. In recent years the imbalance of US trade with other nations has resulted in a pool of overseas dollars which can be tapped for project financing.

Risk identification, evaluation and allocation: Risks need to be identified and evaluated, and decisions made about how they will be covered. The willingness of lenders and investors to finance a project will depend on their assessment of the attendant risk factors; particularly completion risks, operating risks and investment/financial risks. A Compleat Project will be based on a careful allocation of these risks.

Contractual arrangements among project participants: The complex network of contractual arrangement must be developed carefully and methodically to bind the various elements of the project together. These contracts define the roles and responsibilities of the participants, the rewards and penalties and standards of performance. They provide commitments for project construction and operation, the supply of needed fuel and raw materials, and for the purchase of the products. They are also the vehicles for allocating the risks identified above.

Note that none of the three aspects of project development mentioned above relate directly to the design and construction of the project. The processes for designing and constructing facilities have not changed substantially over the years. That is not to say that they have not improved in many important ways – only that the basic concepts and methods continue.

This is in stark contrast to the financing, which has changed so fundamentally that it cannot be compared to earlier financing methods. It is, as they say, 'a whole new ball game'.

Two important interrelated developments have combined forces to bring about these basic changes: One is the growing world economic interdependence and the internationalisation of projects. The other is the shift from corporate financing to project financing.

The concept of project financing has evolved to accommodate the growth in the size and risk of projects beyond the debt capacity of individual corporate, parastatal or government sponsors. Now funds for projects are assembled from a wide variety of sources, often in several different currencies using the project and its revenue – producing potential as the primary credit support for the financing.

1.5 Conclusion

The Compleat Project is increasingly a larger, more complex multinational project utilizing goods, services and financing from worldwide sources. It may employ more sophisticated technologies and methods requiring new realignments of participants, technology transfer and training. At the same time projects must respond to the demand for greater efficiency to compete successfully in the world marketplace for projects. This competition will sometimes pit low-capital-cost, labour-intensive facilities in comparatively low-labour-cost developing countries against high-capital-cost, high-technology, automated facilities in high-labour-cost industrialized countries.

The successful pursuit of world-class projects entails a continuous monitoring of the worldwide market for projects and the international sources of financial resources available to fund them, bearing in mind the fact that, no matter how meritorious a project may be, it will not be built unless the fund can be developed to pay for it.

It also requires an objective evaluation of potential projects to make sure they have the essential features of a Compleat Project; that they meet basic economic needs and yield benefits that exceed their costs; that the major participants are equal to their respective roles in implementing them; that the host country provides an acceptable investment climate; and that the project risks can be successfully managed.

Once these conditions are met, a strategic role in the management of the project development process can impart important impetus toward project implementation. Financing plan development, and risk evaluation and allocation, play critical roles in this process, but these efforts must be carefully planned and co-ordinated with other aspects of project development.

All this requires a high level of management skill and creativity. Each project presents its own set of problems and challenges, but with a high level of resolve applied to their solution these can be mastered, making the Compleat Project a reality.

Chapter 2

Project management: A client's view*

Sir Alistair Frame
Chief Executive, RTZ Group

The RTZ Group spends several hundred million pounds a year in developing new projects, generally in the field of natural resource development, and has a reputation for successful completion of such projects. This success depends in part upon the manner in which the financing of projects is managed and the post-audits that are carried out thereafter.

The RTZ group was one of the pioneers of the discounted-cash-flow technique of project analysis[1] and these are the main techniques used today. A cut-off rate at which a project will be rejected is identified, namely a 10% real (after-tax) rate of return, and for some projects a considerably higher figure is required, depending on the degree of market or political risk. RTZ also analyses markets for the product of the project; investigates the financeability of the project before the project is approved; and carefully checks the interest cover; i.e. the ability of the project to finance debt under very severe conditions such as low metal prices or a reduction in market demand, a most important viability test. The Group also analyses production costs, which one would always wish to be in the lowest quartile of world production costs. Virtually all our projects have been financed by project financing with a debt/equity ratio of between 60 and 70% debt and 40–30% equity. To date, there have been no problems in financing projects in this manner.

The Group has a formal and regular procedure involving a capital investment committee which reviews a project's progress one or two years after commissioning, when the financial and operational performance of the project compared with the original submission is checked. This we find to be a good management discipline.

---

* The lecture upon which this Chapter is based was delivered to the first inter-institutional meeting of the Engineering Project Management Forum held on 19 November 1985 at the IEE, London. The participants in the Forum are the Institutions of Civil, Mechanical, Electrical and Chemical Engineers and the National Economic Development Office.

The procedures adopted by RTZ when developing projects are based on four cardinal rules:

(a) The client must be strong and appoint the best available group (engineer) to execute the design, major supply of hardware and construction on site.

(b) The client and engineer should unambiguously set out the necessary procedures and authorities to carry out the works.

(c) The client should appoint the best available project director to be responsible for the project: he should have clearly designated responsibilities to a Board of directors, probably of a design and project management company owned by the client.

(d) The design, cost estimate and programme must be established and agreed before work starts on the site.

In more detail, the major factors leading to the successful development of a project are, first, a clear definition by the client of the scope of the project. Secondly, careful selection of the engineer and/or the managing contractor and, lastly, a precise definition of the *modus operandi* between the client and the engineer/managing contractor. Each of these three areas is examined in greater detail in the following paragraphs.

In the natural resource field, as in many others, the scope of the project is usually prepared by the client, sometimes with the assistance of an engineering consultant or an engineering contractor. It is also important, indeed essential, to involve senior operating personnel with related experience in drawing up these documents. All too often, the scope of the project is defined by designers and project engineers with little or no experience of plant operation, and this undoubtedly leads to problems.

The scope should include a clear definition of the client's requirements on capacities, layout of plant, ideas on methods of construction, and on the control, size and working conditions of the construction labour force. The client should, at this stage, develop his ideas on his relationship with the to-be-selected managing contractor. Points to be decided include:

Delegation of authorities and methods of presenting change orders for approval from the client, including time constraints on approval

Methods of site contracting and the degree of client involvement in site management and labour relations

Procedures for selection and contracting methods to be adopted on site contracting

Methods of financial control and reporting

Details of meetings that the client requires to have with the contractor or engineer.

Only when this work has been carried out — and even at this stage a considerable amount of money might have been committed — should the client proceed to select the engineer/managing-contractor.

Selection should take place by competition between competent engineers or managing contractors, in consortia or joint ventures with experience in the particular field being developed. The competition should be judged on the experience, quality and availability of the engineer/managing-contractor's staff nominated for the key posts. Experience of the managing contractor in the particular area of the world being considered, and the proposed programme and method of execution of the contract, are also important selection criteria. Finally, there is the method and amount of remuneration, but it should be noted that *all* these factors have to be considered, not just the costs. If this work is carried out carefully and methodically, then the possibility of later misunderstanding between the client and the managing contractor will be much reduced, an area where many projects go 'off the rails'. In evidence to the Public Inquiry into the proposal to build a PWR power station at Sizewell, I emphasised these requirements should be observed in developing the project's organisation, and stressed that, if they were ignored, difficulties may arise at a later stage.

The RTZ Group's method of executing the works is as follows. The client, in establishing the project scope, will have defined a programme of work which will include dates for the completion of preliminary engineering, preparation of the accounting and budget systems, including a cash-draw-down projection, preparation of bid packages, detailed engineering and, finally, preparation of the definitive estimate which will form the reporting basis for financial control.

The engineer or managing contractor will, of course, comment on and often alter the client's suggested programme of key activities in his tender response. The final agreed programme may have to be a compromise between the client's requirements and the managing contractor's resources and ability to obtain delivery of key plant and equipment on the scheduled dates. While in the past it has been the custom to use penalties, more recently the RTZ Group has been using bonuses, providing contractors and engineers with handsome rewards. Nothing talks like money to the contractor.

An integral part of this methodology is the requirement for the engineer or managing contractor to report monthly on:

> Progress in engineering and on purchasing
> Physical progress on site
> Estimate to complete

This latter area is not well done in British project management. If it is done by accountants they may not really understand what is going on: RTZ tend to employ good engineers in this function, who have a sound financial understanding, to make sure that our estimates of 'cost to complete' are as accurate as can be achieved.

The most important part of project management is the people. The client's and the contractor's most difficult problem in the successful development of a project is selecting the best people to manage the activity. RTZ have developed a project management company which employs the key project managers, who are regarded as an elite group moving around the world from one project to another. We realise that we can make or lose more money in managing a project than in any of our operating businesses. This policy has been followed for the last ten years and it has worked. It is essentially to employ the right leaders and to select a balanced, interdisciplinary and business-oriented team, although the rules may have to be somewhat different in the public as opposed to the private sector. This may all sound simple and obvious, but the selection of the top man and his immediate deputies in a project organisation, no matter how small, is probably the most significant element in the successful implementation of a large project. It is an area on which RTZ concentrates.

The most important factor in selecting the top man is his qualities of leadership. Business schools tend to play down the importance of leadership in large corporations and invariably stress management skills rather than leadership. I am sure that exactly the opposite is true in project development, and possibly in business as a whole. Projects in many respects are much more akin to military-type operations than running an operating company, and a military-type *esprit de corps* is almost always required to protect the project from the numerous assaults to which it is always subject. There are not enough top-class project managers with leadership ability available for the work that has to be done anywhere in the world today. As a consequence, it is vital to encourage young engineers to adopt project management as a career.

The structure of the organisation itself is also vital. The project organisation must have a minimum of hierarchical layers; indeed this precept should apply in normal business and industry. In addition, the total number of management, supervisory and support personnel should be kept to the absolute minimum consistent with the workload. From the general manager to the shop floor, in all our normal businesses RTZ have only four levels; it used to be about 15. Wilfred Brown and Elliot Jaques[3-8] were preaching this approach in the late 1940s, but it seems nobody has listened. If the concepts just described are vigorously followed and are not just the objects of lipservice, then the client has a better change of developing a successful project. Whenever RTZ has departed from the concepts described above, the project inevitably goes wrong.

For example, the Group developed the world's largest operating uranium mine in Namibia. The ore body was extremely low grade, 350 p.p.m. of uranium in the ground in an inhospitable environment. Extraction had to be a very large-scale operation to be viable. The scheme was developed with inadequate design data, insufficient pilot plant development, with inadequate engineering and project management and poor financial controls — all because we rushed at it, reacting to our market conception that a window was available for long-term contracts. It was a 'smoking' disaster: the outturn cost was way above budget, the programme was not met and the process did not work. In a space of 15 months, RTZ refinanced the company, changed the management, redesigned sections of the plant and renegotiated the contracts with power utilities in six different countries throughout the world — all in all, a harrowing experience, expensive to the owner, in this case the RTZ shareholders. It happened because we did not methodically go through the process described in this chapter. This happened in 1978, and within two years, the mine was on its way to becoming what it now is — one of the most successful operations in the RTZ Group.

The second example is more recent and in the United Kingdom, although on a much smaller scale. RTZ adopted a fairly new technology, using fluidised-bed combustion for power generation to replace an oil-fired boiler. The project was relatively straightforward, but the technology did not work out as expected. Much more time should have been spent examining the technology, but our engineers were fascinated by being at the forefront of a developing technology — it very rarely pays if you are in business to make money, which is what we in business have to do.

Client organisations, engineering institutions and contracting companies must think carefully about the need for people in this very difficult area. In the whole of the RTZ Group worldwide we have only half a dozen men who are trusted to manage a project of more than £100 million, but there are ten times that number who can run a straightforward operation, whether it is a mine, a factory or an oil plant. The message therefore is that not enough good people make project management their career. Yet it is a very rewarding career. There is nothing quite like seeing something which you have constructed within budget and ahead of time operating well.

2.1 References

Finance
1 SYKES, A. A., and MERRETT, A. J.: 'Capital, budgeting and company finance' (London, 1966)
2 SYKES, A. A., and MERRETT, A. J.: 'Finance and analysis of capital projects', (London, 1963 and 1973)

Organisation
3 BROWN, W., and JACQUES, E., 'The Glacier Project papers' from 'Essays on organisation and management from the Glacier Project Research Series' (London, 1965)
4 BROWN, W., and JACQUES, E.: 'Psychology and organisation'.
5 BROWN, W.: 'Principles of organisation. Part of a Series on Monographs on Higher Management (Manchester 1946)
6 BROWN, W.: 'Some problems of the factory. An analysis of inustrial institutions'. One of a Series of Lectures organised by the Institute of Industrial Administration (IPM London 1952)
7 BROWN, W., and RAPHAEL, W.: 'Managers and Morale' (London, 1948)
8 BROWN, W.: 'Organisation' (London, 1971)

Chapter 3

The project and the community

S. C. Goddard

Director of System Planning, Central Electricity Generating Board

3.1 Introduction

Any large-scale construction works will have considerable effect on the local community and their environment. The Central Electricity Generating Board (CEGB) has a statutory responsibility not only to provide the consumer with an economic and reliable supply of electricity, but also to pay regard to the environmental implications of its works.

Power stations have particular environmental implications and a considerable effect on the local community; they also last a long time. For these reasons, the CEGB has long-standing relationships with the local authorities and with the communities around its stations. The CEGB has therefore adopted a 'Good neighbour policy', aimed at making positive and sympathetic contributions to the local infrastructure and to local facilities.

In interpreting this policy the Board has throughout its existence sought to fit into the community in a harmonious way. However, over the years public and political expectations have grown, and are continuing to grow. We are consequently seeing even more effort being put into those aspects of projects which have attracted local attention in the past, and more consultation aimed at identifying areas of concern and measures to deal with them.

In order to ensure that the Board's policies regarding the community are effective, and to marry successfully the planning engineers' and designers' concepts to the project managers' discipline of constructing to time and cost, it is apparent that, from the beginning, close co-ordination is essential as the project proposals are developed.

3.2 Project stages

To many of those involved in the construction industry the granting of planning consent marks the birth of a project. However, an increasing amount of work is necessary to reach this stage, with gestation often lasting several years.

In the case of construction of power stations in England and Wales, the CEGB is required to apply to the Secretary of State for Energy for consent under the Electric Lighting Act (1909) as amended. When granted, this consent carries with it deemed planning permission.

The process leading to the granting of consent can be divided into several stages. First, site identification and proving; this is likely to take at least two years. Secondly, detailed planning and negotiations with the local authorities other bodies and interested parties. It is during these periods that the major effort is needed to assess the impact of the project on the community and to work up proposals for alleviating measures. Finally, the Secretary of State for Energy may call a Public Inquiry to assist him in reaching a decision on the application for consent.

Inevitably the overall planning process takes many years, possibly as long as the construction phase (typically seven years). During this time circumstances may vary, requiring changes in emphases, and, of course, public attitudes may alter along with perceptions of the project and the environment implications.

3.3 Consultation with local authorities

It is a matter of fact, from observation and experience, that there is a growing concern for the environment. Thus it is increasingly necessary to look deeper into the effects of new projects on local communities and to assess what measures should be taken to avoid, reduce or ameliorate disturbance. This can best be done in close consultation with County and District Councils, parishes and local people.

Whilst the local response to proposals to construct a power station vary from one area to another, there are a number of typical considerations which will always need to be taken into account when drawing up detailed plans. The main common local issues are:[1]

(a) Prospects for local employment on the site, for local supply and service industries and for local businesses

(b) Worries about road traffic, not only the overall increase that the new project will cause but the types of vehicle and peak times; road improvements to deal with increased traffic flows

(c) Concern about accommodation for migrant workers and the disruption they may cause to welfare and social facilities

(d) Anxiety about loss of rural amenities, visual intrusion and effects on wildlife.

Whilst the emissions from both nuclear and fossil-fuelled stations do cause a great deal of national and international debate, the experience of the CEGB is that these feature rather less in the local negotiations than might be expected. The limits on gaseous and liquid emissions from power stations are usually set by statutory licencing bodies, such as the Ministry of Agriculture, Fisheries & Food and the Health & Safety Executive. Frequently these limits rely on a philosophy known as ALARA (As Low As Reasonably Achievable).

In drawing this balance, the planning engineer must fully understand the needs of the project manager and his ability to honour commitments entered into. Where necessary, such commitments can then be included in contract enquiry documents sent to tenderers.

Local authorities have a significant role to fulfil in looking after their interests and identifying measures beneficial to local people. However, in addition to their obligations to the electorate, these bodies have statutory duties as the planning authorities. These two responsibilities may at times be in conflict, and local people may expect a narrow response from council members and officers, limited to protecting the immediate locality, rather than one benefiting the population as a whole.

3.4 Public appreciation

People who live in the area around the site for a proposed power station are likely to see a need for all sorts of measures to negate the impact on their lives, e.g. new roads and bypasses, undergrounding of transmission connections, extensive landscaping, etc., which in themselves could disturb other people. Money spent on such measures would come from the pockets of the consumers, including local residents, and a careful balance must be reached on how much money should be spent and which measures will produce the greatest overall benefit.

Not only may the public see a need for spending considerable sums of money to protect the local environment, but different people have different and possibly conflicting views on the ways it should be done. Of course, everyone may agree that a new road is required, but each person will also believe that it should be away from them. For coastal sites some people may argue for a new harbour to enable materials to be delivered by sea or for a new railway line; others may say 'not on this stretch of coast' or 'not across this beautiful countryside'.

In order to complement its own work to identify both real and perceived issues, the CEGB has funded a number of studies, notably by a team at Oxford Polytechnic and by the Department of Psychology at Surrey University. The effects imposed on local communities by the construction and operation of nuclear and coal-fired power stations have been surveyed, the important issues identified, and the interactions with the local community analysed.[1,2]

Surprisingly the research has shown that, whilst some aspects will undoubtedly have unfavourable effects on the community, the overall conclusion is that power stations can be absorbed into a rural area with relatively limited impact on local conditions, bearing in mind the scale of the projects. One of the major uses of the research has been in providing factual information on which to base future proposals and to refute some of the more exaggerated claims of objectors. Perhaps some of the most interesting recent research has been into the potential economic effects on the local community when Trawsfynydd Power Station in North Wales eventually reaches the end of its life and has to close.[3]

In summary, this research has indicated that the eventual closure of Trawsfynydd Power Station, which is located in a thinly populated rural area, would result in serious economic and social impacts. These can be divided into four main areas:

(a) Loss of direct employment. A number of the present employees would not be willing to transfer to other posts with the CEGB
(b) Loss of indirect employment
(c) Loss of future employment prospects. At present 70% of vacancies at the station are filled by local residents.
(d) Loss of potential in-migration. The remaining 30% of vacancies are filled by persons moving into the area. These provide a valuable contribution to the social and cultural mix.

In order to reach a considered view on proposals for a new power station and on ameliorative measures to be adopted, it is essential to ensure public involvement during the project planning stages. When the CEGB's interest in a site is announced, some people may react immediately while others may see any firm development as being far into the future, if at all. All too often at this formative stage the CEGB is accused of being secretive because it cannot give firm details of the proposals for the project. As time passes and no construction plant appears, other than the investigation borehole rig, the initial interest may wear off.

The CEGB aims to provide comprehensive information on proposed projects to the public and to encourage local people to comment on its proposals; the Board is also seeking ways of developing its consultation procedure in this area. At Sizewell the Board for several years has operated an information centre on the site and information shops in nearby Leiston and Ipswich. The public can call in and view exhibitions, collect information, and ask questions about the project. This approach, an extension of the CEGB's practice of temporary or mobile exhibitions, has been invaluable in assessing pubic interest and in presenting a more human face to local people.

Another means by which public opinion can be gauged is by local meetings. However CEGB experience is that the tone of meetings is often rapidly set by the most vociforous, whilst many local people present may feel too intimidated to speak, either to ask for information or to give their support. Add to this the courage needed by speakers to face a possibly hostile audience, and it can be seen that the alternative of a one-to-one discussion at an information centre or exhibition is often preferable. Local meetings are perhaps most useful when the audience are invited local-interest groups.

3.5 Assessing the balance

In negotiating with the planning authorities and other bodies the CEGB has to reach a considered and balanced vew on benefits to be gained and the cost of certain actions. The CEGB has statutory obligations towards the provision of economic and secure supplies of electricity and towards protection of the environment. These duties can introduce potential conflicts. The CEGB's judgment on the balance between them is wide open to public scrutiny, and is accordingly made in a political context. The satisfactory achievement of the Board's objectives depends to a large degree on public acceptance of the CEGB's activities and behaviour. It is important, therefore, that the Board adopts and is seen to adopt a sensible balanced approach.[4]

A number of specific considerations follow:

3.5.1 Prospects for employment

The chances for improved employment and local expenditure arising from the project, both during construction and operation of a power station, can be of major benefit, leading to a degree of acceptance in the area.

Until recent years the general practice in the heavy construction industry was to employ mostly travelling men who would move from site to site. Current practice, however at least for unskilled and semi-skilled labour, is to recruit more locally. With long-term projects such as power stations there is considerable potential for employment, training and retraining, as the construction progresses, of local labour. In addition, main contractors are encouraged to use local suppliers and sub-contractors wherever possible within the bounds of good commercial practice. At Sizewell the CEGB is establishing a directory of such local suppliers and sub-contractors, and will encourage all the main contractors to use these companies. However, contractors must retain sovereignty over their work in order to ensure completion to time and cost.

3.5.2 Traffic

For convenience, reliability, economy and versatility, road transport of construction materials will usually be preferred. The preferred approach of a developer would therefore be to opt for road delivery and pay for the highway improvements that may be agreed with the local authority. However, this approach ignores the evident public concern about heavy traffic even on major roads. Power-station sites tend to be in predominantly rural areas, and the delivery of construction materials by road may have a significant impact on the locality with occasional peaks of around 200 movements a day (i.e. 100 deliveries and 100 return journeys) of commercial vehicles. To this of course must be added the workforce travelling to and from the site, usually in private cars.

With the real and emotional impact of such road-traffic movements in mind, the CEGB must take all reasonable steps to minimise traffic and its effects without unduly prejudicing the construction costs or timescale.

The alternatives to road transport, i.e. rail and sea, for delivery of materials, and the practicality of these will vary from site to site. One of the most beneficial actions can be to transfer from road, to rail or sea, the delivery of bulk materials such as aggregate, of which some half a million tonnes or more may be required. Unless there is already a good rail access to the site and provision can be made for reception facilities without major environmental consequences, for coastal and estuarine sites, the most advantageous method of delivery could be by sea, with the material being pumped ashore by pipeline. This, however, is not a perfect solution which can be universally applied. Deliveries would be weather dependent, facilities must be provided for washing and storing, not only increasing the site area required, but introducing problems of possible saline contamination of local water courses and agricultural land. Nevertheless the CEGB sees this approach as having a considerable net benefit, at some locations, to the local community in removing large numbers of heavy loads from the roads, but its use will be dependent of the nature of the coast and frequency of difficult sea conditions.

Sea delivery of other materials may well involve the provision of some form of jetty or harbour, and their construction could have unacceptable environmental and/or recreational consequences. Even then, anything other than an all-weather harbour, navigable at all states of the tide, would introduce what most construction managers would consider to be an unacceptable constraint.

With reduction in the numbers of manufacturers and suppliers, works having direct-rail-access double handling is a problem with rail transport. If there is no railway onto the site itself, materials would have to be transferred over local roads by a 'shuttle service' from the railhead, introducing problems in its own right. Extending the railway to the site can be costly and difficult. At many sites such provision would be so environmentally detrimental as to create its own objections.

3.5.3 Social disturbance

As mentioned earlier, there has been a general trend in the UK away from the employment of large numbers of migrant unskilled workers. To a great extent this may have removed such problems as existed around major construction sites in the past. Nevertheless the public are still apprehensive about the prospects of a 'drunken mob' rampaging through peaceful villages. Local memories sometime exaggerate any problems that may have been caused by construction workforces in the past, and there is a tendency to place the blame for any incident on a migrant group, who will form a convenient scapegoat.

A review of the construction phase of the Sizewell A Power Station (1960–65) has shown, however, that in this case there was in reality very little difficulty caused by the workforce. Whilst there was an increase in petty crime in the area, much of this was in fact due to theft from the site. There was perhaps an increase in drink-related misconduct, but, no doubt surprisingly, and certainly contrary to local memories, police records show that the incidence of sex offences actually dropped during construction and rose thereafter.[4]

Nevertheless the public understandably remain sceptical. Any influx into an established community, for whatever reason, will to some extent cause disruption and resentment amongst the local people. If the public are to accept a major construction project, with its associated workforce, then steps must be taken to ensure that the disruption, and hence the resentment, is kept to a minimum.

There are several approaches to this problem. Maximising the employment of local people will reduce the size of the incoming workforce. That element of the workforce whose permanent homes are outside reasonable travelling distances can be encouraged to live, during the week, in a residential construction village on, or close to, the site. The practice of the CEGB, where a construction village is provided, is to ensure that the accommodation is of a high standard with good-quality dining and recreational facilities etc. Lists of potential lodgings may be prepared, and often caravan sites will also be made available.

An important way of minimising the total disruption is to reduce the project overall construction period. This can be greatly helped by working 'double day shifts', i.e. morning and afternoon shifts, from Monday to Friday. Weekend working is limited, enabling the non-local workforce to travel home for weekends, thus further avoiding the risk of local conflicts.

3.5.4 *Loss of rural amenities etc.*
A large power station unavoidably occupies a considerable area of land, and, together with its associated transmission connections, is likely to be visually intrusive in the local landscape. In the UK much of the countryside, and even more of the coastline, is highly valued and protected, either on ecological or aesthetic ground and often for both. It is possible that a power-station site will encroach on one or more protected area, and a transmission line can be an alien presence in scenic areas.

Unfortunately it is imposible to avoid some ecological impact arising from the construction and operation of a power station, nor can the works be made invisible. In the detailed investigations and planning of a site, environmental studies are initiated to identify species of flora and fauna of interest, and, if particularly rare or valuable specimens are present, wherever possible to propose ways to avoid undue disturbance and/or restore them on site or nearby. Measures to achieve this may, for example, include the establishment of 'seed banks' or nurseries to maintain stocks of rare plants.

In order to mitigate the visual effects of power stations as far as is reasonably practical, the CEGB appoints eminent architects and landscape architects, who will have a major input into the planning and design of a project from an early stage. In the past, site landscaping has been limited to land in the Board's ownership, usually within the boundary of the site. However, recently moves have been made to adopt a more liberal approach and, with the active support of the local authorities, the CEGB has entered into agreements with local landowners to provide and maintain specified tree and shrub planting to screen the power station from particularly sensitive viewpoints. At a time when woodlands, hedgerows and copses are under considerable threat from modern farming measures, this approach will make a positive contribution to the conservation of the countryside, in addition to meeting its primary purpose.

3.6 Positive benefits

Up to now this Chapter has only considered ways in which specific effects of construction can be mitigated. However, the public and the local authorities are beginning to expect developers to put forward steps to provide specific proposals to compensate an area for the inconvenience and disruption which a large-scale project may cause. The argument that no amount of amelioratory measures will result in a zero impact and that other measures should be undertaken to balance the impact appear to be gaining ground. Counter to that argument are the questions as to whether, in the case of power stations, the electricity consumer should pay, and how much, and also what measures are appropriate.

The identification of appropriate recipients for such benefits cannot rest solely with the developer, although he may decide on the limit of funding. The treatment of local communities and the establishment of objective criteria for any planning gain or compensation around different power-station sites must be seen to be fair and not based on the principle that 'he who shouts loudest receives the most'.

3.6.1 Economic compensation

The system of rating (local taxation) in England and Wales is such that, whilst the ESI contributes a considerable sum of money to the local authorities (totalling about £340 million a year), under the present Rate Support Grant formula a comparable amount is removed from the grants payable from Central Government funds. There is therefore little or no net financial benefit to local government, and this does not seem to be equitable.

In some countries the electricity authorities make specific financial contributions and taxes to local municipalities during the construction and operation of power stations which they retain. In France, EdF (the national electricity utility) make large payments. At the Golfech site in south-west France a fixed payment of FFr 10M (£1M) a year is being made during construction, and FFr 6M (£0.6M) a year will be paid during operation. Italy have adopted a similar approach; ENEL having indicated that the municipality of Timo, near the PWR site at Piedmont, will receive approximately £10M during construction and £2.5M a year once the station is operating.

In Japan the local communities around power stations benefit from national government grants and local property taxes. The grants, financed from taxation of the electricity utilities and aimed at improving local services and infrastructure, range from £4M–£12M for a 1000 MW fossil-fired plant, to £21M for a 1000 MW nuclear plant. The local property taxes produce revenue over an assumed lifetimes of 15 years by means of a formula which produces, again for a 1000 MW nuclear station, about £7.6M in the first year, decreasing to £2M in the last. Compensation to individuals is also paid in Japan. Fishing rights are purchased from fishermen whose catches may be adversely affected, and a regional co-operation fund compensates for psychological anxiety or problems arising from the presence of a nearby power plant.

3.6.2 Infrastructure improvements
In the UK the philosophy has generally been that the developer of major works should pay for such improvements to the local infrastructure that are made necessary by the construction and operation of the works itself. In the past, the CEGB has paid for road improvements and other transport facilities, new sewage works etc., and has assisted towards the provision of improved health, education and housing where these would otherwise be overloaded as a result of an increase in population caused by an influx of construction staff and their families.

As an alternative to the cash compensation paid by some countries, there is the option of financing specific projects in the area, e.g. where recreational facilities may be limited or overwhelmed by an increase in population. These could be of direct benefit to the construction workforce and to the developer's staff, such as a swimming pool or other similar facility, and have a long-term benefit to the community.

Such a concept does, of course, have its attractions but also drawbacks. For example, part of the population may prefer a swimming pool, others may prefer a community centre or a substantial contribution towards the restoration of the local church. Reaching agreement may not be easy.

3.7 Conclusions

In a small densely populated country like the UK all of man's activities affect the lives of his neighbours to a greater or lesser extent. The construction and operation of large projects such as power stations, whether they be nuclear or fossil-fired, can impose major effects on the character of the area. It is important, for good community relations, that project-planning engineers and managers should do all within their means to encourage local acceptance. Proposals for measures to ameliorate the effects of construction can best be assessed and decided in close consultation with local authorities and the local people.

The greatest difficulty faced by a public utility such as the CEGB in promoting and maintaining acceptance of a project and good community relations is to assess the balance which must be achieved between expenditure and the benefits to be obtained. Given sufficient resources it would be possible to redress the grievances of the majority of local objectors, but, of course, the ultimate source of the finance would from the public themselves.

3.8 References

1 GLASSON, J., *et al.*: 'A comparison of the social and economic effects of power stations on their localities'. Power Station Impacts Research Team (Oxford Polytechnic, 1982)

2 LEE, T., *et al.*: 'Results of public attitude surveys towards nuclear power stations conducted in five counties of South West England'. Department of Psychology, University of Surrey

3 LEWIS, P. M.: 'The economic impact of closure, without replacement, of Trawsfynydd Power Station'. Institute of Economic Research, University College of North Wales, Bangor, May 1985

4 GAMMON, K. M.: 'Evidence to Sizewell B Pubic Inquiry: Transcripts Day 286 (13 Nov. 1984)' Department of Energy, 1984, p. 71

5 GAMMON, K. M.: 'Environmental Implications of Power Generation for the available options'. Council for Environment Science and Engineering, Oct. 1982

Chapter 4

Some thoughts on high budget projects

A. J. O'Connor

General Manager, The New Brunswick Electric Power Commission

4.1 Prologue

Large, high-technology projects are full of uncertainty and risk, and especially so at their inception. Their very size creates uncertainty; new technology multiplies it. A wide variety of necessary project ingredients such as regulatory agencies, community relations, changing political demands and pressures, inflation, timely delivery of engineering and hardware, and the availability of competent labour further compound the uncertainty. But it is management's job to make decisions in the face of uncertainty. Lord Halifax recognised this in 1693 when he said, 'He that leaveth nothing to chance will do few things ill, but he will do very few things'. and what he said then is even truer today. Management in general must deal with ever increasing uncertainty, and project managers particularly must do so.

When schedules slip and cost estimates grow, it is perfectly normal to question the project management's ability to manage. While the major reason for such changes in the direction of a project has been the many changes imposed by uncontrollable forces, this perception has been unwittingly nurtured by our tendency to characterise projects by single-valued measures — a single cost, a single start-up date — which gives illusions of certainty. The economist, Ken Boulding, said, 'An important source of bad decisions is illusions of certainty'.

The basic function of management is to make decisions. A decision is an irrevocable commitment of resources, where resources are time, money or manpower. Note the word irrevocable. Unless the commitment cannot be retrieved, it is not a decision. An outcome is the result of committing resources. We make decisions, but we are judged on our outcomes. Outcomes can either be good (desirable) or bad (undesirable), or they can be intended or unintended. We usually cannot be certain of good outcomes because we cannot completely control forthcoming events. Therefore, we must try to enhance the probability of good outcomes by making good decisions.

A good decision is simply one that is carefully evaluated within the available time with a defined logic. It is consistent with the values and information available at the time to the decision maker. Where there is uncertainty as to which events might occur, the logic of the decision process should include that information. Making decisions in the absence of uncertainty is a trivial process for which management personnel are not needed.

At the inception of a project, quantitative standards are established as to materials, manpower, money and time, all for the purpose of measuring accomplishments. This is done with the full expectation that forthcoming events will require substantial modification of the bases upon which the project will be measured. When the external world is highly uncertain and continually changing, this means that the basic measurements standards for the project management system are continually changing. Therefore, there is no constant guidepost by which to judge the success or failure of the project, and the outside observer receives the illusion that the project has grown out of control and the project management system has failed.

4.2 Background

Although the ingredients of project complexity have been found to be fundamentally the same for various types of mega projects, for purposes of discussion it is perhaps beneficial to deal with only one. The construction of nuclear power plants has become the most costly, complex, and risk-laden area of involvement undertaken by North American electric utilities. Until the late 1960s, the cost of new electric generation was predictable and acceptable. The risk factor was relatively low for a utility building new facilities; technology was proven, inflation was nominal, and capital was readily available and inexpensive. Completed power plants were mostly fossil-fuelled and the economies of scale of the rapidly increasing size of the generation units had kept the cost per kilowatt of capacity from rising sharply. In most areas, load growth had been regular and predictable. Power plant construction was a reasonably orderly, stable activity that did not attract a significant level of public attention.

The late 1960s and early 1970s brought the advent of construction of nuclear generating capacity in quantity. With this came new technology, substantially higher construction costs, and greatly extended construction schedules. Thus, utilities were required to raise increasing amounts of capital for investment in projects whose return was further in the future. These new and unfamiliar challenges were recognised as additional risks to be undertaken; however, the economies of nuclear generation, the substantially lower projected cost per kilowatt hour of output warranted then, and still warrants, the assumption of this risks. Many utilities, in both Canada and the United States, were seriously considering, if not proceeding with, nuclear power projects.

The early and middle 1970s saw the addition of other complicating factors particularly applicable to power-plant construction. Probably the most significant of these was rapid inflation with its effect on the decision-making processes of nuclear project managers. Of near equal importance was the nationwide shortage of competent engineers and designers necessary to produce the specifications and drawings in the time spans required to support nuclear project schedules. Added to these changing events was the management-diverting fact that nuclear generation was making the electric utility highly visible in the community as a result of widespread public concern about the possibility of dangers associated with commercial use of nuclear power.

In the late 1970s and early 1980s, some of the more detailed reviews of the nuclear project-management problems concluded that there was no simple way of resolving the major difficulties in this field. In many cases, the project were disadvantaged through organisational processes which, although they had proven adequate on many previous hydro and fossil projects, were not able to cope with the unique and onerous demands of nuclear design and construction. The most prominent, and no doubt the most important, of the organisational deficiencies of many projects undertaken in that era was the absence of the owner's corporate experience in dealing with the many urgent and potent decisions required, and, in some cases, the tendency of the owner to delegate this authority. Expressed another way, it has been found that bringing technical and project experience into an organisation for employment on a nuclear project, or on any large project for that matter, does not, in itself, provide adequate response to the problems which arise.

4.3 The proponent

Clearly, no one is more inclined to give a project the priority it requires than the owner who will have to live with the result for the foreseeable future. Increasing demands in areas such as finance, siting and licensing dictate that the owner must not delegate the overall decision-making process.

The extent of the owner's sustainable direct project involvement depends, of course, upon the corporate expertise available within the organisation. It may be that this should be his first area of examination and, if found in need, strengthened. Although management structure is not, in our view, as important as commitment and dedication, it is of significance. A mega project, and perhaps particularly a nuclear project, will require many corporate decisions to be made promptly and decisively. Corporate preparedness is, therefore, important. Although there is no precise method of replicating successful project management because of its great dependence upon the characteristics of so many, nevertheless the approach to the problem will influence the outcome. We would, therefore, suggest that the first step towards the development of a project's management should be to enhance, if necessary, the owner's corporate capability with respect to very large projects.

Our second suggestion would be to avoid total reliance on the traditional pyramid organisational form.

Senior-level management has found it difficult to maintain a properly supportive and helpful approach to a giant project when it is informed solely through a pyramid-organisational leader. When the returned span of attention is so broad and the incidence of decision requirements so frequent, project managers have found it difficult to fully advise and utilise the support available to them. It is therefore recommended that a performance audit function be established reporting to a management committee.

The purpose of the performance audit function would not be to evaluate decisions made by line management, the excellence or otherwise of which will become apparent through the overall project progress, but rather to expose the project to the scrutiny of highly experienced personnel who are in a position to evaluate whether or not the various parts of the project organisation are adequately equipped and manned, and are in fact performing according to expectations. This audit group may also be in a position to anticipate and advise on potential construction problems before they occur, thus exposing the management committee and the project management organisation to an alternative consideration which would not be available without close access to the work by highly qualified experts who are free from daily responsibilities of execution and administration.

It must be recognised, of course, that performance audit is a concept requiring a high level of tact and diplomacy if it is to be constructive and helpful to the project. On the other hand, project managers have become accustomed to financial-audit reporting to the next level of responsibility, and, we are confident, would in a short time find performance audit to be certainly no more debilitating. In the final analysis, project managers should find aid and assistance, together with an element of security, in the performance audit system. It could be argued that, if such a system had been in place within the American space programme, the Shuttle would still be flying, or conversely it would be recognised that the program as executed contained a calculated risk as opposed to a failure of management communications.

This general management philosophy, which gives the project manager direct access to a management committee, results in the very best and most extensive experience and expertise being made available to the project.

Additional areas wherein the owner's presence should be exercised would include:

> The owner, at the full corporate level, should consider project concerns as a high-priority matter

> The owner should assert himself with all contributors

The owner should insist on being satisfied with regard to the necessity of demands or suggestions of the regulators.

The owner should remain directly involved to the degree that his expertise and resources will allow

The owner should retain the care, custody and control of the project

The owner should ensure that planning and engineering are well advanced before commencement in the field

The owner should spare no effort to ensure that the licensing process is advanced as far as physically possible before commencement in the field. Broad statements of confidence on the part of the regulator that the project will at some point meet all requirements should not give rise to unreasonable optimism

The owner should anticipate the possibility that he may be required to dedicate additional resources to the project.

However, a large complex project requires someone to plan, organise, staff, evaluate, direct, control and lead it from point of inception through commissioning to commercial operation. This is defined as the basic role of a project manager. The need to assign one man to be responsible for achieving the ultimate goal of the project, i.e., commercial production at economical cost, is equivalent to the need to assign a single chief executive officer of a corporation.

The project manager alone, however, cannot satisfy the above requirements, and his responsibilities are usually delegated to individuals best suited for the given task because of their skill and/or experience. When talking about the Project Manager's Office, one is talking about many qualified individuals joined together for the common objective to establish and maintain control of the project with a view to assuring its completion on schedule and within budget while achieving the desired technical proficiency and overall safety. Quite often, project management, or the pulling together of all necessary participants in the correct sequence at the proper time to produce a flawless result, can be compared to conducting an orchestra of a thousand musicians and not producing a single sour note. Conducting the orchestra is probably easier as the musicians have a common goal, whereas contributors to the project process have widely diverse motivations.

Needless to say, the number of qualified individuals available to an owner to perform the duties of a project manager and other key positions on a large complex project are very limited. It is most important that the persons filling these positions have the necessary qualifications, experience and proven records of performance on previous work of similar type and magnitude.

The project manager should be appointed at an early stage of the project and empowered to take all necessary action to ensure the success of the project. This results in the project being planned and controlled on an integrated basis;

i.e. the project will have a single point of integrated responsibility, the project manager, who will be supported first of all by a dedicated project team and additionally by the project management Committee.

4.4 Contracting

Traditionally, up to the early 1970s, owners of many projects relied on large design–build firms. These firms were engaged under firm-price contracts because of the continuing access to 'state-of-the-art' technologies and qualified staff. The availability of firm-price prime contracts wherein the contractor assumed most of the risk of construction enhanced this approach. During the 1970s, firm-price contracts for work on mega projects became unrealistic because of construction uncertainties.

There were many factors which brought about the failure of this historically successful method of contracting; however, the underlying cause was the fact that many contractors grossly underestimated their costs. The reasons for this included lack of experience in the particular type of work being bid, failure to anticipate the impact of newly imposed environmental and regulatory requirements, and misjudging labour productivity. The success of this method of contracting also suffered owing to the long time periods required for performance of the work, with some nuclear projects extending up to 12 years duration.

Whatever the reasons, when a contractor committed to a firm price finds himself in a loss-bound position, he quickly reacts to protect his financial interests. He will search very carefully for any external events, influences or situations outside his control which have, in reality or otherwise, adversely affected his operations. This search inevitably produces a blizzard of claims for more money and extensions of time, many of which may be highly imaginative. Some may have merit, however, and this fact requires that all claims be processed and adjudicated. This process consumes precious time during which management's attentions is diverted to an essentially non-productive task, while the work is being executed at a rate considerably less than full speed.

The result of all this, in our view, is that, if a mega project is undertaken on the basis of prices established through firm price bidding, the basis of that decision may well change. In other words, even though the project would appear to be protected by the contracting system, in actuality it may not. When the output of each of the project participants is closely bound to the project of a preceding activity by others, flaws are certain to appear. This is especially true of mega projects because of the proliferation of preceding participants and activities, some of which have no project responsibility whatever. In addition, the consequences of non-performance become very large indeed. They can easily far exceed the corporate worth of almost any contractor.

Consequently, from a firm-price contract point of view, a condition can easily arise where the non-performance of a participant is clouded by the impact of numerous other contributors, none of which can be expected to do a perfect job all of the time. Under these circumstances there is considerable danger that a conflict will develop, resulting in useless litigation, from the owner's point of view, while large impediments to problem solving are put in place.

As a result of these circumstances, we believe that firm-price contracts are practical only in instances where the design is essentially complete at the time of tender and where significant design changes are not anticipated. On nuclear power projects, these prerequisites are generally not attainable. The nuclear steam-supply system is subject to continuous application of new standards or the introduction of new methods of verification of old standards by the regulator. We have concluded that the preferred approach to nuclear-power-plant construction is to engage, at the outset, on a cost-reimbursable basis, a general contractor as a construction manager. This contractor, in addition to managing all construction work, would engage sub-contractors to perform specific packages of work on a firm-price basis wherever the status of design, material availability and site access warrants this approach. All other work would be performed on a direct-hire basis.

A cost-plus contract of this nature effectively shifts the contractual risk associated both with actual construction and with other external uncertainties, such as regulatory and inflation, to the owner. The owner pays essentially all project costs plus a fee to the contractor for his services. This fee may be calculated on a variety of bases.

Because of the size and complexity of this type of contract, it is common for two or more contractors bidding such work to join together in a joint venture for purposes of the specific contract being considered. This allows the member companies to combine their entrepreneurial talents, technical and financial abilities, equipment and staff, thus enabling organisations to pursue work for which individually they are not prepared to commit all the skills of capacity necessary for the contract.

From an owner's point of view in awarding a contract to a joint venture, one should ensure that one of the joint venturers is identified as the 'sponsor', so as to allow it to operate as a single entity so far as the owner is concerned. The sponsoring organisation should be in a position to demonstrate full contract scope experience both from a proposed staff and corporate point of view. To facilitate corporate commitment on the part of the members, their names should appear in the name of the joint venture. This makes the individual firms more sensitive to the success or failure of the project.

In order to enhance the probability of a cost-reimbursable type of contract being successful, the owner and the contractor should each recognise and accept that a relationship of trust and confidence has been established between them. When problems arise, the parties must be prepared to agree on the means to overcome them and mount the necessary resources needed to solve the difficulty.

While the owner must be prepared to accept the increased risk associated with developing a major project, he looks to the contractor for specialised knowledge in planning and executing the construction effort, including the provision of supervisory and trades staff. The owner may also look to the contractor to provide all or a portion of the systems capability necessary to monitor and report project progress, although some owners may elect to use their own systems which they are comfortable with and which have been proven on previous projects.

Since firm-price contracts must be marked up by the contractor to cover the risk associated with the uncertainties he anticipates, their cost is somewhat high. If the uncontrollable event occurs, the contractor remains whole; if it does not, the cost to the owner is higher than the cost of the work. If the cost of the uncertainty is higher than that allowed by the contractor in his estimate of risk, the project will probably find itself either approving an extra or involved in litigation.

Although cost-reimbursable contracts provide the least risk to the contractor and the least apparent protection for the owner, they are, under appropriate conditions, the best buy. This type of contract requires, however, the greatest project-management and contract-administration effort by the owner to ensure satisfactory project completion at a reasonable cost. Dollars saved in contractor mark-up should be applied to increased management effort. Some owners advocate the use of incentives for contractors as a means to encourage optimum productivity and cost-effectiveness. One type of contractor incentive used by some owners involved putting specific portions of the contractor's fee at risk, based on his success in achieving predetermined cost and/or schedule milestones. If these specific milestones are not met by specific dates, the previously identified sums are not paid to the contractor. Some contracts go even further by including provision for a bonus/penalty fee whereby a specific amount is identified to be paid to the contractor for each day a major (critical) milestone is achieved in advanced of a scheduled date; conversely, a specific daily amount is deducted from amounts to be paid to the contractor for each day required after a scheduled date in achieving a major milestone.

Some owners oppose contractor incentive schemes because of the feeling that they result in an adversarial climate similar to that which often accompanies firm-price contracts, brought about by the difficulty to agree on the impact of the inevitable project changes which occur. These owners feel that the relationship of trust and confidence is better served by the absence of incentives, and replaced with a greater degree of contract administration by the owner to ensure optimum contractor productivity and cost effectiveness.

On a typical cost-reimbursable contract, the contractor accepts an obligation to plan and execute the work for a fee, with the owner paying essentially 100% of the costs. The contractor's obligation has a substantial price tag associated with it when measured in terms of cost to the owner for equipment and materials to be properly handled by the contractor. In addition, the cost includes the labour for installation according to specification in a cost-effective manner. On the other hand, the contractor will probably have no assets at risk and only minimal exposure to the possibility of an overrun in his fee allowance for home-office senior management and executive staff.

It could be asked whether or not contracts as related to mega projects are becoming obsolete. Certainly what we might call traditional forms of contract do not always suit the special and complex relationship which must exist between the owner of a major project and his contractors. It is not practical or of any value for such an owner to insist that contractors bear obligations which they will not be able to meet. This fact has a profound effect on contractual relationships as well as management style. Contracts, however, must not be considered unimportant; rather, owners must take an active role in their structuring and administration to maximise control over project performance and minimise the risk they have assumed. Organisational and contractual approaches are highly interdependent, and are the primary means by control is pursued.

4.5 Trust and confidence

It may be that the proper establishment of a condition of trust and confidence between the owner and each of his major participants, as well as between them and their suppliers, should be a major thrust of an owner contemplating a mega project. Confidence is not necessarily established through either the highest or lowest price. It is perhaps useful to recall that price and cost are not the same thing, especially if one is looking at the lowest of a number of prices. The price is the anticipated cost of commencement, and can change if forthcoming events are not as originally contemplated. Confidence is probably more properly approached through an analysis of previous experience and performance. Experience applies to both the corporation offering the service and to the individuals being proposed to execute the various functions. They are of at least equal importance. Confidence may be a shared emotion. The owner must, in our view, establish his confidence in the contractor, but so also should the contractor have confidence that the owner will in his turn respond appropriately to exigencies of the work. Some would say that the contractor should be kept comfortable, where 'comfortable' is defined as the absence of stark fear for his corporate life. Perhaps more realistically the contractor should have confidence that the owner will in fact execute those responsibilites he has undertaken to deliver or have delivered, that he will commit additional resources if they are required, and that he will be assertive with all participants and insist that they achieve an acceptable performance or remove themselves from the project. Trust is perhaps of a different flavour. Trust is required partly because, in its absence, constructive problem solving becomes laborious if not impossible. Trust could be an absence of apprehension that the other side will unfairly interpret conditions or exaggerate minor oversights or omissions. Trust could be the establishment of a worthy relationship acceptable to both parties at the corporate and personal levels.

The establishment of a condition of trust and confidence between the owner and his contractors is then a matter of importance and one worthy of a considerable effort and investment on their part. Trust and confidence may also be, to a degree, the product of a determined effort from project inception to minimise conflict both between the project participants and the owner, and between the project and external groups. Although some may claim that conflict can be constructive, when it is associated with a mega project where there is a large number of participants and hence a large potential for conflict, it is, without doubt, destructive, and has serious cost connotations for the owner. Time and effort spent in developing good contractual relationships as well as well defined contract scopes is, therefore, a very good investment.

4.6 Labour relations

Good labour relations and the establishment of a harmonious work climate on the site are extremely important, especially for large projects requiring peak forces of several hundred tradesmen belonging to numerous trade unions. We believe project agreements which control working conditions and rates for the life of the project, and isolate the project site from labour disputes which might occur in the surrounding areas, are necessary. These agreements should also establish the basic criteria for the resolution of disputes in their early stages, and assist in monitoring labour stability.

The objective of a project agreement is not to extract promises of an unusual nature from either side, but rather to create an agreement which will allow labour to know in advance that all crafts will work under standardised starting, stopping and other work rules. The agreements should also provide grievance procedures designed to accommodate jursidictional differences and other conflicts between the trades. Such agreements identify the rights of both labour and management.

Regarding hiring practices, the construction manager should act as the only employer on the site, and all trades personnel should go through a central hiring office and payroll system. Individual contractors and/or sub-contractors, however, should control their own men and be responsible for their performance support their costs. The payroll costs and relative charges under this approach would be met by the individual contractors and sub-contractors.

Actually it is suggested that the construction manager establish and operate an employment office on behalf of the various site contractors. Since it is expected that a project of the size of a nuclear development would be a unionised situation staffed through the various trade-union hiring halls, it is suggested that very deleterious influences can infiltrate a project if any particular union hall is doing business with more than one entity on any one project. It is therefore being suggested that the various site contractors should requisition their trades through an employment office operated by the construction manager on behalf of all site contractors. Such a system would provide project-wide management of hirings and firings as well as uniform enforceable labour administration. Personnel hired through the system would become the employees of particular contractors, thus preserving the direct relationship between them and their management.

4.7 Supporting the project

Mega projects, and perhaps especially nuclear projects, require large numbers of contributors, all of which should benefit from the project. The list would include manufacturers, designers, constructors, labour, financial institutions etc. A project also requires an owner, or several owners on occasion, all of whom should benefit from the project. In addition, a project must be located somewhere, and it should also benefit that area through an increase in the commercial activity, the provision of a tax base, etc. In other words, a substantial gain should result for all parties, both those directly involved and those indirectly involved.

Perhaps one of the difficulties experienced by most mega projects, with their large array of participants together with those indirectly impacted, has been an absence of a willingness, or a recognition of the need, to facilitate the project as well as to benefit from it. There existed, not so long ago, a belief on the part of project proponents that the project could support all demands put upon it from whatever source. This philosophy has been shown to be seriously in error. Future mega projects will need to find a way of enlisting the support of all project benefactors including the local areas, labour, suppliers, government and the regulators. They will need to do this without any relaxation of real safety or environmental standards.

4.8 Conclusions

In conclusion, it may be said that an understanding of the dynamics of mega project management exists in many quarters. Information systems have been developed capable of arranging and classifying data so that the managers can be in full possession of all related facts on a continuous basis. It could be said that management by exception is an antiquated concept, and that management by total information is now possible. This concept enhances the possibility of improving on original estimates should circumstances permit, rather than being always satisfied if original objectives are achieved. The problems of organisation, leadership and commitment are better understood than perhaps they were only a few years ago.

It would appear, then, that most project specific needs are quite capable of acceptable execution. The area of continuing concern relates to developing a system through which the needs of the regulators can be provided to the project during its normal design cycle, and the development of the conviction on the part of the many benefactors of a project that their active support is required if they wish to sustain economic progress and enhance their current situation.

Chapter 5

*Management of joint venture projects

C. Fleming
Manager, Construction Division, Projects Department, BP International Ltd.

5.1 Introduction

Joint venture or multiclient projects have become a normal fact of life in the oil industry and in other capital-intensive industries where the capital and resources required, and the risks taken, in embarking on and executing a major project have become so large that few, if any, companies can undertake such projects without a partner.

This joint venture effect leads to many conflicts which pose fascinating, if not unique, problems in the execution of such projects; all of this is of course in addition to the not inconsiderable difficulties of managing successfully projects which are large and technically complex.

This Chapter deals with the nature of large joint venture capital projects and the factors affecting their management.

5.2 Nature of joint ventures

In the oil industry it is normal for one participant (usually a major company with the appropriate resources and expertise) to be designated as 'Operator' to act on behalf of the others.

Taking an oil field as an example, the Operator is responsible for the exploration, drilling, development of project concept, execution of the agreed project phase, and operating the field and plants after completion. Project management is thus one aspect of being 'Operator', and must be satisfactory to all the participants.

Clearly there can be a wide range of participants taking part in a joint venture for a wide variety of reasons. All will be in it to make money, but there will be

* The author wishes to thank BP International Ltd. for permission to publish this Chapter.

be many additional or subsidiary reasons. Some participants will be major companies maintaining or extending their expertise; some will be companies diversifying or gaining experience at relatively small risk; some will be small companies who may see the project as an investment opportunity only, or as a chance to develop taking advantage of the expertise of the major participants; sometimes a company may be willing to accept lower than normal returns to obtain access to a country's natural resources − oil, gas, minerals etc. Participation in a joint venture spreads the risk and frees capital for investment in other attractive investments. Also sharing facilities may be more economic − common terminal, common pipeline etc.

Many countries no longer permit foreign companies to own 100% of a project, or even a majority shareholding; thus a company may have to combine with local or national companies in the host country. In addition, host government participation may have to be accepted. Such government participation may be for reasons of trade or prestige, or to control or influence the development of a resource, or to provide employment in the chosen geographical regions. The point to recognise is that economic viability may not be the criterion for government participation.

Sometimes joint venture arrangements occur as a result of 'natural forces'. In developing an oil or gas field, say in the North Sea, an Operator may find that the field extends into adjacent licence blocks held by others. The government allocating the licences will be concerned to ensure that rational development of a national resource takes place and requires the companies involved to make joint agreements to develop the field effectively and economically. This is known as field unitisation and, if the licence holders do not produce an acceptable scheme, government reserves the right to impose a solution. To date in the North Sea this has not been necessary since the licence holders have always managed to make acceptable agreements based on factors such as the distribution of estimated reserves, the difficulty and/or cost of extracting the oil or gas from the various parts of the field, etc.

Frequently, participating companies and perhaps government will regard the 'Operator' with suspicion. The project manager must overcome this and gain their trust and respect, while learning to live with delays in decision making (committee effect) and the legitimate desire of each participant to be informed fully, and the less legitimate but understandable desire to exercise maximum control. However, a participant is entitled to feel that his interests are receiving due attention and that the project team are performing effectively and giving value for money.

While the companies taking part in a joint venture formed for the execution of a particular project may have a common interest in achieving its aims, there can be considerable differences of emphases because of the interaction of other activities in which many of the partners will be engaged − often in the same type of business and possibly even in competing projects. The reconciliation of these different interests is difficult and will be a key factor in the success of such a project in realising the normal objectives of operational effectiveness within budget and schedule. An example of this conflict might occur where one participant is anxious to generate a return from his investment as rapidly as possible, while another participant wishes to extract maximum longer-term benefit which might mean a slower, possibly more elaborate, development. Another frequent area of difficulty for the project manager is the different priority, perhaps the changing priority, which each participant accords the project start date.

The shareholding sought by individual participants will depend upon many factors — availability of capital, desire for market share, desire to control the project either commercially or politically, whether securing management of the project is an aim, i.e. technical control, etc.

Four examples from BP's wide participation in joint ventures have been selected to illustrate the range of shareholding and management options as follows:

Sullom Voe Terminal Project, Shetlands
BP: operator/project manager
BP: minority share 7% during construction phase
Ula Project, Norwegian North Sea
BP: operator/project manager
BP: majority share 57.5%
North-West Shelf Development Project, Australia
BP: not operator
BP: minority share 16.67%
Middelburg Coal Mine Project, South Africa
BP: not operator
BP: majority share 87%

Details of the overall participation in these projects are given in Appendix 5.8. It should be noted that being Operator with a very small holding (Sullom Voe) and not being Operator with a very large holding (Middelburg) are very unusual situations.

5.3 Project management of joint ventures

The nature of joint ventures has many implications for the management of them. The project manager no longer represents a sole owner, but is in a position of stewardship for all the participants. He is much more involved in the politics associated with satisfying other companies, and sometimes government.

The project manager may report to a management or executive committee of the participants while retaining a functional reporting relationship with his own management. Ensuring that the participants are satisfied, or possibly accepting a compromise between their needs and those of the Operator, may heighten the conflict which often exists in a major company matrix organisation between achieving the relatively narrow aims of a particular project and meeting the wider corporate interests. Thus the project manager may well be subjected to significant additional stresses in a joint venture situation.

Alternatively the Operator may find it expedient to appoint a project director to deal with all matters relating to the participants, the management committee and possibly government. In this case the project manager, who then reports to the project director, is relieved of much of the political involvement and is able to concentrate on managing the project more or less as he would when working for his own company as sole owner.

However, considerable extra reporting and presentation work are still required. Also, participants will want to audit all activities on a regular basis, and more justification of actions and decisions may well be necessary. Thus the project activities are subject to much more outside scrutiny, and actions have to be taken thinking, 'How will X see this?' All participants must be treated impartially and must perceive that this is so; e.g. there should be no preferential treatment for the Operator because he is the project manager's employer.

A management support group and effective information systems are essential because of the need for good presentation, provision of timely information, and consistency. Participants must be 'carried' with the Operator. An autocratic style is not possible or politic — diplomacy is essential.

Some type of AFE (approval for expenditure) control will be needed. It is important to structure this realistically so that AFEs reflect the work/contract breakdown. This improves control and lessens the work of cost allocation.

The Operator may not have all the resources needed and may wish to use contract/agency staff, but participants may claim that they are paying for Operator expertise not staff 'bought from the street'. Also participants may be concerned that the Operator is not providing his high-calibre staff for their project.

Participants may wish to second staff, and this should be accepted providing it is clear that they work only for the project team and do not report separately to their parent companies. This is an important principle which should be applied rigorously. Such secondees are, of course, quite distinct from the team which a non-operator participant may appoint to look after his interest. If this representative team is close to the Operator and has authority, it can be of great benefit to the project in ensuring speedy decisions and approvals from a participant whose head office may well be remote from the project.

Relationships between the project management and the contractors may be more complicated in a joint venture because of interference or influence by the other participants.

Contractors should resist this and the project manager should insist that they do so even though they may find it difficult in practice.

Participants may make or appear to make unreasonable demands for information or justification. Project staff must not be distracted from the main thrust of achieving project objectives. Personalities are important in establishing the right relationships.

High-level contacts between the company who is Operator and the other participants will occur. Those which may be related to many interests, as well as applying to the specific project, need careful handling; this may well be the kind of help which the project manager can do without.

The project manager must have the respect of all. Not only must he be in control, but must be seen to be.

5.4. Types of project management

Bearing in mind the nature of joint ventures and the implications for managing them, it is interesting to consider various arrangements which might be adopted. A summary of these is given below; on balance the first, using a majority participant with 'know-how' and resources, is best.

5.4.1 Operator is a majority participant who has 'know-how' and resources
This provides advantages of cohesion and control by one company. The protection of the Operator's interest automatically provides effective protection for the other participants. However, the Operator, with his complete knowledge and with the majority shareholding, is very much in control and may 'bulldoze' the other participants. No Operator would admit to such tactics, but the temptation is strong.

5.4.2 Operator is a task force from all participants
This is theoretically attractive but presents real difficulties in forging an effective team from different backgrounds working with unfamiliar standards and procedures.

5.4.3 Operator is a minority participant
This prevents 'bulldozing' by the Operator, but would he have the resources required, or be willing to use them on this project?

5.4.4 Operator is an independent professional project manager
This places reliance entirely on professional integrity and skill. While this should be sufficient, it is not reassuring for participants.

5.5 Agreements

It is essential to have a clear joint-venture agreement covering the obligations and relationships between participants. The agreement should set out the voting procedures to be adopted, including whether vetoes may apply, and whether the majority required to carry a vote will vary according to the subject of the vote; e.g. the commitment to proceeding with the project may have to be unanimous.

Audit provision and the role of any management committee to control the joint venture should be defined. Also the agreement normally formalises the appointment of the operator and gives the framework within which he will manage the project, including what rights of access participants have to operator and the rules for seconding their personnel, etc. Sometimes a separate agreement is drawn up to define the role and responsibility of the Operator.

If a joint venture format is adopted the companies involved will be participants and not partners; i.e. no participant has liability for the others and each would expect to fund his share in his own way. Alternatively, a corporate body could be formed, which would have completely different implications and liabilities for the shareholding companies.

It should be remembered that the joint venture agreement is not just a legal agreement; it is intended to create a framework which will permit the project to progress without delay, albeit under control. Accordingly, those involved in drawing up the agreement should understand the nature of joint ventures and the project management requirements; i.e. engineers and project managers as well as lawyers and accountants must contribute. Otherwise, unrealistic or restrictive, or possibly counter-productive, controls may be imposed on the operator.

5.6 Conclusion

Clearly joint ventures are here to stay because of the advantages of sharing the capital outlay, spread of risk, investment opportunity, etc. From a project management point of view it is tempting to say that there are no advantages — only problems. This overstates the position of course, since very often participants can assist materially. For example, they might have particular expertise or be able to provide valuable contacts etc. The project manager must not close his mind to such potential benefits, and in fact has a duty to seek and exploit them.

In forming the joint venture the decision on which company is to be operator/ project-manager is crucial. Although the percentage shareholding may be a factor, it is vital to select a company having the resources, expertise and experience necessary to implement the project sucessfully.

In a joint venture, the Operator has full responsibility for the implementation of the project and, with his executive authority and depth of knowledge, enjoys a very large degree of control in practice, albeit subject to the constraints and checks imposed by the other participants.

It is important for the Operator as project manager to decide how to deal with the other participants, how they should be informed, how they are to contribute, and how to encourage and harness their involvement while avoiding undue interference. The Operator has to accept more reporting and presentation, and more justification of his actions and decisions. There is no doubt that the key to a successful project is establishing a relationship of mutual trust and respect between the project management team and all the participants.

The Operator's attitude and freedom to act will depend to some extent on his shareholding. If it is a majority then he can be more autocratic, but, of course, other constraints may apply; e.g. he may be in other joint ventures with some or all of the same participants, sometimes with the roles reversed. If the Operator has a minority shareholding, then more interference has to be accepted. It is unlikely that a major company would be keen to have the Operatorship if his shareholding was less than 15%.

If a participant is not Operator his degree of control and influence will depend largely on the extent of his shareholding. If a majority, the participant effectively controls the funds, and indirectly the project. Although the Operator still has the responsibility for implementation, the team appointed by the majority interest will have great authority and will be able to exert strong pressures on the operator. If a minority shareholder, the participant has to decide whether to be active or passive. If his shareholding is less than 10% he must accept being overruled or outvoted. If it is 15% or more he should expect a veto, in which case a balance has to be struck between exerting influence sufficient to protect the interest, and interfering unduly with the Operator. In this respect a minority participant must guard against being blamed for causing delays, or being blackmailed into precipitate action.

Finally, experience has demonstrated that, unwieldy and inefficient as they may appear, joint venture projects can be implemented successfully when managed with an appreciation and anticipation of the difficulties and frustrations inherent in them.

5.7 References

The following may be consulted for other views and advice on joint ventures:

1 'Check List: Establishing a Joint Venture' (Export Group for the Process Industries)
2 WEARNE, S. H.: 'Principles of engineering organization' (Arnold, 1973) chap. 6 (drawing on BIM conference The Management of Consortia for Large Projects, London, 1970, and earlier publications listed)
3 'Responsibilities for project control during construction'. Report TMR 17, Technological Management, University of Bradford, 1984 (see Appendix C for a job specification for the role of project director)

5.8 Appendix: Examples of participation in joint venture projects

Sullom Voe Terminal Project, Shetland: Operator − BP

> 33 participants with ownership proportions ranging from 0.03% to 19.5%
> BP share during construction was 7%
> This increased to approximately 15.7% during operation

Ula Project, Norwegian North Sea: Operator − BP

BP	Conoco	Statoil	Pelican	Svenska Petroleum
57.5%	10%	12.5%	5%	15%

North West Shelf Development Project, Australia:
Operator − Woodside Petroleum

	Woodside Petroleum	BP	Chevron	Shell	BHP	Mitsui/ Mitsubishi
Domgas Project	50%	16.67%	16.67%	8.33%	8.33%	N/A
LNG Project	16.67%	16.67%	16.67%	16.67%	16.67%	16.67%

Middleburg Coal Mine Project, South Africa:
Operator − Rand Mines

BP	Kahn	Rand Mines
88.503%	6.497%	5.00%

In listing the participants and their percentage holdings, shortened generic company titles such as BP, Shell, Conoco, Kahn, etc. have been used for simplicity of illustration. It is felt that giving the full corporate titles of the participating companies is unnecessary in the context of the issues being considered, and would have detracted from the clarity of presentation.

Chapter 6

Contract strategy

F. Griffiths
F. Griffiths Associates

6.1 Introduction

Commercial projects are usually launched with a profit motive. To be successful the project or service needs to be better, more fashionable, cheaper or more economical than the competition at the time the product or service arrives on the market, and must remain so over the planned market life.

Major commercial projects include factories, oil fields, mines, office and shop development, air terminals, bridges, tunnels, or could be a product such as a new aircraft, vehicle or ship, sometimes together with a new factory to manufacture it. On a lesser scale, the overhaul or refurbishing of the schemes mentioned above can still be significant projects.

The programmed time of a major project is measured in many months, and maybe years. All too frequently, over such periods the market changes.

The risks facing the entrepreneur are daunting. Even if the product is right in concept, the estimates are good enough and the market does not change for the worse, inadequate project planning and execution can still steal profitability: cost or time overruns or failure to meet the design specification in any way can turn a potentially successful project into a disaster by opening the market to competition or simply because the increase in project cost makes a project or service too expensive.

It is therefore vital that the project manager and his staff fully understand and are committed to the cost, time and performance objectives of a sanctioned project.

6.2 Estimating

> Estimating is not an exact science. Hindsight may give 20/20 vision — even then, one has to look!

The estimating problems of the entrepreneur launching a new project are illustrated by the Channel Link, for which proposals were made to the French

and British Governments at the end of October 1985, and the choice made in January 1986. Whilst earlier Channel tunnel proposals were expected to be funded by the governments, the schemes required 'Every penny to be raised by the promoters at their risk'. The estimated cost schemes varied between £2 billion and £6 billion; the construction period between 3 and 6 years, and two of the passenger traffic passenger traffic forecasts were:

Estimate	Vehicles per year	
	1993	2003
Channel Tunnel	25m	29m
Euro Route	27m	36m

The profitability of the selected scheme will depend upon:

    Traffic volume
    Charge per vehicle
    Cost of construction
    Time of construction
    Interest rates
    Exchange rates
    Inflation rates
    Depreciation rate
    Maintenance costs
    Service and operational costs.

and not least:

Competition from the ferries.

The Channel Tunnel scheme was selected by the governments, apparently on the grounds that it employed existing technology and was seen to involve the minimum cost in real terms. (Other schemes required technical development and were thought to be under-estimated). The risk to the governments of financial failure of the consortium was thus minimal. Further, a tunnel could be abandoned, whilst it would be difficult for government to leave a partly completed bridge if the consortium did fail.

It is considered that a sanctioned scheme should therefore remain viable for a foreseeable variance from the mean estimated and should particularly take account of the cost of the risks that the owner finds it necessary or desirable to carry himself.

There are numerous examples worldwide where a project failed to meet the commercial objectives. Ultimate commercial success depends upon the market for the product or service being at least as good as was foreseen over the projected life. It also depends upon the project being constructed within budget programme and specification; these are the issues which are addressed by project and contract strategies.

The following brief descriptions of project failures illustrate how dramatic failures can be; they also show that, whilst high- technology projects are particularly susceptible to cost and time overruns, conventional projects where the technology is well established can also suffer unforseen delays and cost increases.[1]

Private company failures tend to remain private unless they result in the failure of the company. For example, little would have been heard of the RB 211 problems if Rolls Royce had not gone into receivership.[1] Failures attracting most publicity are usually those financed by public money.[2] Some of the most dramatic financial failures include:

> The building of five nuclear power stations in Washington State, said to be the most expensive publicly financed construction in history. The total cost of the five nuclear stations, which the Washington Public Power Supply System (WPPSS) − popularly known as 'Whoops' − decided to build in the early 1970s rose from an original estimate of less than $4bn to more than $100bn. By comparison the Space Shuttle programme cost $8bn.
>
> The Sydney Opera House which took 17 years to build and cost ten times the estimate.

UK examples are:

> The Humber Bridge cost 3 times the estimate and took double the project programme
>
> The Carsington Dam and Tunnel, where, *before the Dam failed* in June 1984, the reported delay was 90 weeks and the final cost was estimated as three times the budget
>
> The Concorde which cost £1134m. against the original estimate of £160m.

These cases demonstrate some of the major problems, thereby pointing to aspects of project launch needing particular attention:

> The ease of under-estimating costs, especially in areas of new technology
>
> The difficulty of writing complete specifications
>
> The problem of estimating time, especially for developing new technology
>
> The optimism of designers, especially when seeking financial backing
>
> The optimism of project managers when estimating their ability to recover a situation
>
> The difficulty of cancelling a project once started, even when the estimated cost to completion continues to rise
>
> The need for a project and contract strategy which gives the owner the best chance of success, and sufficient information and control should problems arise.

Shakespeare foresaw most of the problems:

> When we mean to build, we first survey the plot, then draw the model;
> and when we see the figure of the house, then we must rate the cost of
> the erection; which, if we find outweighs ability, what do we then, but
> draw anew the model in fewer offices, or at least desist to build at all?
> Much more in this great work — should we survey the plot of situation
> and the model, consent upon a sure foundation, question surveyors,
> know our own estate, how able such a work to undergo, to weigh against
> his opposite; or else, we fortify in paper and in figures, using the names
> of men instead of men like one that draws the model of a house beyond
> his power to build it; who, half through, gives o'er and leaves his part
> created cost, a naked subject to the weeping clouds and waste for churlish
> winter's tyranny.
> HENRY IV Part 2, Act 1, Scene 3

6.3 Project management and contract strategy

A successful project not only requires a successful analysis and estimate but also
successful strategy and management. A contract strategy should depend on
many factors, not least the resources available to the owner. He needs a
professional project manager, able to provide overall management of the project
with the owner's interests at the front of his mind.[3] He will need design
resources, civil, mechanical, electrical, chemical or electrical. He may need
manufacturers, suppliers, contractors and commissioning staff. Those that are
not available 'in house' must be brought in.

Contract strategy deals with the division of the project into separate contracts, the timing of contract awards, the form of the contract most likely to
encourage satisfactory completion whilst providing controls and opportunities
to the owner to rectify problems before they cause serious difficulty to the
project.

6.4 Risk sharing

The contract establishes the risks to be carried by each party. The general
principle suggested is that risks should be carried by the party best able to either:

(*a*) control the risk, or
(*b*) estimate the risk

When the risk is to be carried by the contractor, the extent of the risk carried
may be limited to that which he may reasonably be expected to control or
estimate, taking into account the custom and practice of the industry of which
the contractor(s) form a part.[4]

Contractors should be expected to include a contingency for the unforeseen risks.

It is for the employer to so design his inquiry documents so that risks are seen by both parties, and hence that the contingencies included are the best estimates available.[5] The employer should also recognise the risk areas he has decided to carry himself and should include in his project estimates to cover them.

The designer of the contract document should always keep in mind the primary objective of the project, which is usually to build to specified performance within the programme and budget. Hence, the contracts should incorporate agreements on how to minimise and control the significant risks, identify who is to carry such risk and to what extent, and should attempt to include adequate incentives to meet the employer's objectives, preferably by attempting to align the objectives of the employer and the contractor.

An outline contract spectrum is given in Appendix 6.8.

6.5 Matters influencing the type and conditions of contract chosen

(a) *Resources available from suitable contractors*: Managerial, technical, plant, labour, financial and other resources required. Before placing any contract one has to be satisfied that the contractor will be able to marshal the resources required. The availability of resources to the contractors in your field may limit the types of contract selected. Particularly check the financial standing of the company with whom you intend to contract. Remember, many 'famous name' companies have subsidiaries with limited liability. Account must be taken of the custom and practice in the industry of which the contractor forms a part.

(b) *Resources available to the employer*: A small employer with few resources tends to be pushed towards lump-sum single contract projects or to use consultants or management contractors on a fee basis. Larger employers with their own design and contract management teams can move further down the spectrum. Project managers should take into account the certainty with which his organisation's resources will be available to them when needed, in selecting type of contract.

(c) *Multi-project contracting*: An employer rarely in the market may seek to impose conditions which could rebound on a multi-project employer. A contractor may seek to take maximum advantage of a single-opportunity employer.

(d) *Complexity of project*: The number of industries, companies, the number of trades and the number of trades unions necessarily involved.

(e) *Size of project*: Can adequate competition be obtained for the extent of work if a single contractor is appointed for a large section of the works? Can the contractor supply an adequate proportion of the labour from proved permanently employed men?

Where the project requires the construction of more than one unit, consider obtaining qualifications for the entire project and alternatives for similar number of units. The project may then be launched on a minimum scale with the option of releasing further units at predetermined stages if the initial work proves successful.

(*f*) *Time available*: Should the programme be optimised to minimise the commissioned cost including interest during construction? Do market considerations justify increasing the cost of the project in the expectation of greater overall returns?

Remember to allow a time contingency for each contract and for the overall project which takes account of the risks of delay accepted under the contract (e.g. provisions for extension of time).

(*g*) *State of the design*: Is the design a prototype, development, repeat of previous design not yet commissioned, repeat of established design, etc.?

Consider releasing design and development work in stages, so as to better retain control of potential cost overruns.

(*h*) *Likelihood of design variation*: At the inquiry stage there is a tendency to be optimistic that few design changes will follow. In projects involving complex plant, operational experience is likely to point to where economy, safety or reliability can be improved. Variations on the grounds of economy are difficult to justify once the project is launched, since they usually involve a programme risk, and hence cost, which it is easy to under-estimate. It is, however, prudent to include in the contract a means of establishing the change in cost to the employer and making variations within a limit which probabilities suggest may arise.

(*j*) *Number of contracts into which the project is divided*: The project may be divided into as many contracts and sub-contracts as there are areas of expertise required. Indeed, some projects require more than one contractor for each area of expertise. On the other hand, a turnkey or comprehensive contract would involve a single main contract. However, use of a single contractor may only delegate to him the problem which the employer would otherwise face. The decision is influenced by many of the other considerations discussed.

The NEDO Large Industrial Sites Report[6] and Guide to the Placing of Major Contracts[7] recommended reducing the number of employers on a large industrial construction site to the minimum; e.g. to limit the number of site employers to the number of unions involved. NEDO argued that this would enable contractors to attract and afford high-quality managers, and that their small number would allow ready communication between them. Experience has not always demonstrated a successful outcome, and this has been particularly noticeable where the labour force of any one employer exceeds 500.

(*i*) *Single contract* — *sometimes known as comprehensive, package deal or turnkey*: Delegates procurement, manufacturer's installation and commissioning to a managing contractor. May include some or all of the design. This offers the simplest solution to an owner, who need only provide a specification. It demands the minimum contract administration resources from the employer.

The specification might identify the product, the rate and efficacy of production required, the design life, the operational costs, the maintenance costs and the target capital cost (may include interest during construction and cash flows.)

Assessment of tenders for comprehensive contracts can be complex and require significant technical and commercial expertise and judgment; e.g. the operational costs can only be demonstrated after a significant operating period, by which time the employer's only recourse is to damages.

If the employer has the technical ability there is thus a tendency for him to limit his risk by specifying some of his technical requirements in detail. But this can lead to commercial difficulties; e.g. it may be that the contractor can demonstrate that individual technical requirements are not compatible with the overall specification. Serious difficulties arise if the employer seeks to vary the design after work has commenced — this can involve substantial delay and payment for abortive work, additional software, additional hardware, time-related and financing costs, possibly to a greater degree than if the employer has more detailed control of the project.

(ii) *Multi-contract projects*: The employer divides the work into areas of expertise, e.g. design manufacture, 'civil', 'electrical', 'mechanical', or 'supply and delivery', and 'installation' in the expectation that the economies exceed the risk. He may let contracts as sufficient information becomes available in an attempt to save time ('fast track' construction[8]).

The main risk is that the employer must take the responsibility for the effects of failure of one contractor on another and for the effects of his own failure on each contractor. Risks increase sharply with the number of contractual interfaces; i.e. usually the number of contractors employed with problems of one late contractor affecting succeeding contractors. Thus work should be divided so that the information and access interfaces are minimised. There is usually a 'Pyramid' organisation of sub-contractors. If this results in too many employers on site, it could be avoided by having erection contractors for material supplied by others. The responsibility for the satisfactory operation of such material must be identified in the contract.

(iii) *Design contracts*: A major problem on advanced-design large industrial sites has been to finalise the design of plant, and thus, for example, of the foundations required to support it. There is therefore a place for plant design contracts undertaken before site work begins. Design time is relatively cheap when no hardware is being made. Variations are similarly cheap at that stage.

6.5.1 Problems

How does one recognise that a design contract has been completed? What provisions should be made if the design information on which other work is based is subsequently varied due to the designer's responsibility?

(k) *Information and access interfaces*: The contractor will attempt to judge his risk arising from delays in information and access from the employer, the engineer, or indirectly from other contractors, and the effect this may have on his cash flow and on his need to claim extra from the employer.

(*l*) *Size of the contract relative to the financial resources of the contractor*: Paying stage by stage as pieces of work are completed can increase the number of contractors available, but reduces the incentive to complete, compared with paying only on contract completion.

(*m*) *Industrial relations*: Contractors' employees are concerned over:

   (i) Continuity of work
   (ii) The degree of certainty of payments
   (iii) Increasing the real value of their income.

Uncertain site bonus arrangements often lead to problems. 'Motivation', working conditions, site facilities and safety precautions are of increasing importance.

Employers and unions in a single industry do not necessarily adopt a uniform attitude on the same site. Employers and unions may not necessarily take the same view from one site to another. More than one union may represent a particular trade. The views of all parties change with time and circumstances.

Some contractors assert than in some areas of the country the output per man is so uncertain that they would be reluctant to commit themselves to lump sum or remeasure contracts.

The National Agreement for the Engineering Construction Industry reached in 1981 between the employers and their trade unions has brought a degree of consistency into the wage-payment systems of large industrial sites. Coupled with the long standing Working Rule Agreement for the Civil Engineering Industry and the Joint Industry Board for the Electrical Contracting Industry, prospects for the management of industrial relations on large sites are somewhat improved, even if construction demand increases.

(*n*) *Risks outside the control of both the contractor and the employer*: These include abnormal weather conditions, unforeseen ground conditions, technical difficulties which could not have been foreseen at time of tender, national strikes, changes in statute − safety, occupational, environmental.

(*p*) *Conditions of contract*: A wide range of conditions of contract published by professional Trade Associations and major employers are available. The intention of those published by professional bodies was to divide the risk between the parties. However, the most recent UK editions (in 1985) were published before 1980 after a long period of a sales market, and some employers feel that the balance of risk had moved too far against the owner. Conditions of contract used overseas (often based on UK forms) now tend to favour the purchaser and recognise the importance of international trade by providing clauses dealing with financing and currency problems. The major UK conditions of contract are currently being reviewed.

Production of 'In-house' conditions or the modification of a well known set should be approached with care. It is sometimes easy to change a clause without recognising the consequence on others. In any event, a change to an established set of conditions of contract forces a contractor to assess the change in risk to him during the often too short tender period, probably resulting in additional contingencies or qualification.[9]

Conditions of contract from suppliers are often presented in 'legalese', printed in small grey type on the reverse of the quotation. The theory may be that 'that which is difficult to read and understand will not be read'. However, it has not been unusual for such conditions to include clauses which when put in simple English mean:

> The delivery date is given for indicative purposes only
> The price to be paid will be that ruling the company's price list on time day of despatch, whenever that happens to be
> The Vendor gives no warranty that the goods will be fit for the purpose intended by the Buyer
> In the case of a dispute between the buyer and the vendor, the vendor will be the sole arbiter

The threat of such clauses to timely completion to budget can hardly be over-emphasised.

The areas in which discretion is most freqently sought are those which influence the incentives to the contractor:

(i) Terms of payment; e.g. single payment, progress or stage payments, budgetary control payments, milestone or keydate payments (see below)

(ii) Liquidated damages and bonus payments related to one or more dates of the contract, extent of damages or bonus, period over which the damages or bonus arrangements are to apply, liquidated damages (or bonus/penalty arrangements) can also be applied to other vital areas of the contract, e.g. rate of production, efficiency and availability.

The Department of Transport has recently introduced a successful incentive for accelerating work on motorways. Tenderers are required to allow in their bids for the rent of each lane at £x000 per mile per day; i.e. the tender consists of the estimated cost of carrying out the work, plust the estimated lane rental based on the contractor's estimate for the time taken to complete the works. Thus early completion effectively gives the contractor a substantial bonus whilst delay requires the payment of additional lane rental charges. As a result, successful contractors bring far more plant to motorway repair projects, resulting in much earlier completion.

Dr. Martin Barnes has suggested that site rental charges could be extended generally to commercial projects now that contractors are willing to carry the risk. The site rental charge could equate to the financing cost of the site as it develops.[10]

Another recent development has been to require the contractor to work to a detailed programme rather than to an overall completion date. It is argued that the absorption of 'float' on a non-critical programme reduces the probability of the completion of the contract programme, since the critical path cannot finally be identified until the contract is nearly complete. Damages or the withholding of a progress payment may be used as an incentive to maintain the detailed programme. This arrangement not only encourages early completion but requires both parties to manage the contract properly on a day-to-day basis. After initial reluctance, most contractors have welcomed the 'key date' procedure, since it has also encouraged the management of projects within the company.

The Model Conditions of Contract recommended by UK professional institutions have required the contractor to meet the specified requirements at time of takeover and to repair or replace any defect during the defects liability period without limit to the cost incurred.

Conditions of contract sought by suppliers, particularly overseas, often seek inclusion of total limit of liability clauses, which, if accepted, could prevent the employer from obtaining his specified requirements.

In conditions of market uncertainty employers may find it advantageous to include a clause for 'termination at the employer's convenience'. However, it should be recognised that contractors are likely to take account of the consequences of early termination of an important contract. In cost reimbursable contracts the circumstances under which the contract may be terminated by the employer should be given particular attention.

It is possible to seek competitive tenders for target-cost contracts, but clearly the normal rules of competitive tenders do not automatically apply.

Generally it is concluded that there is no more reliable way of letting a tender within budget and to programme than by inviting competitive tenders, especially when well prepared inquiry documents and specifications are available.

6.5.2 Warranty periods

It is important to state the design life required of the plant where design responsibility rests with the contractor.

Warranties sought should obviously have regard to the design life, and perhaps to the increasing difficulties the owner may have as time passes to prove that a defect has arisen from a design rather than maloperation. In general, it should be borne in mind that the law has greatly strengthened the position of the individual consumer under the consumer protection acts; e.g. the purchaser of a washing machine may well have legal redress for defects over a ten-year period. Industrial buyers are not so protected because the law deems that they are able to take care of themselves. However, a contract using a facility specification with a stated design life with no other reference to warranty puts the buyer in a powerful legal position if he can find a supplier to accept it. Usually some compromise is necessary, and the buyer often feels best protected when his quality requirements are fully met.

(*q*) *Negotiation or competitive tenders*: Lump sum and remeasure contracts tend to be difficult to negotiate with a single tenderer because:

(i) At the outset the employer's side have an estimate of the expected cost. Unless the design has been finalised, it is likely to be below a contractor's estimate since it is all too easy to leave some item of expense out of the estimate and because Project Departments tend to be optimistic in the hope that they can launch their particular projects.

(ii) Contractor's ability or wish to quote a competitive price in negotiation is influenced by:

> His position in the market at the time (his price varies according to his work load and the marginal resources available to him), and
> he may endeavour to take advantage of his negotiating position in that once selected, the employer may have insufficient time to open negotiations elsewhere.

A very high degree of confidence between the parties is needed, both for contract administration and to refine future estimates. The principle should be 'equality of information'.[11] The client should be able to monitor performance, particularly where there is a cost-plus element. Unless a very high degree of confidence has been established with the contractor, and particularly where time is of the essence, 'selective competitive tendering' is the most certain means of letting a contract to programme. In cases of uncertainty a target-cost contract can be let by competition.[12] Whilst target-cost contracts can be very successful, it should be recognised that contractors may perceive that a greater profit can be made by persuading the purchaser to increase his target rather than using the contractor's management effort to minimise the cost.

There is often benefit in developing specifications in co-operation with potential contractors, and here a 2-stage competitive tendering process is recommended. Tenderers should be advised of the intention to revise the specification after the tenders have been submitted and that a similar number of tenderers will be selected to bid against the revised specification.

This form of post tender negotiation can be both technically and commercially successful and expeditious. Other forms of post-tender negotiation should be approached with caution and be carefully documented, if on the buyer's initiative.

(*r*) *Government legislation*: Government legislation may affect the profitability of a contract by changing:

(i) The physical design and works (usually for safety or environmental reasons)
(ii) The way in which the works are carried out, e.g. change of scaffolding regulations
(iii) The provisions of the Finance Act, e.g. changes in Corporation Tax, SET, VAT
(iv) Other legislation, e.g. Inflation-control legislation.

Most conditions of contract provide for the additional costs of (i) and (ii) to be met by the client, although in many contracts the real cost would be difficult to establish.

During the 1960s and 1970s successive Governments introduced a variety of prices and incomes legislation, and yet there were also substantial periods when there was no statute in force. This practice introduced further hazards for tenderers; e.g. in one period a statutory 'wage freeze' led to substantial payment of bonuses which could not be recovered under escalation clauses. On another occasion a 'productivity' deduction of 50% was introduced, which was intended to prevent the cost of wage increases being carried into prices. Obviously during a period of high inflation real productivity savings of that order could not have been anticipated — therefore introducing a further risk to tenderers. On another occasion the widespread introduction of artificial productivity bonuses again raised problems for estimators. On a long-term project the real uncertainty for tenderers is that the government may do something during the contract period which will influence the profitability of the contract for which they are bidding.

(s) *Price variation clauses or contract price adjustment*: Price-adjustment clauses have been long sought by manufacturers and contractors to protect them on long-term contracts from the effects of inflation. The advent of computers allowed quite complex systems to be administered so as to follow the apparent change in the value of money with reference to particular products or services.

Most clients resist appeals for 100% price escalation, mainly on the grounds that the contractor should be left with some encouragement to resist inflation himself. However, UK productivity failed to increase at the same rate as in competing industrial countries, and in some cases it noticeably worsened during the 20 years ending about 1980.

The recession following the 1979 oil-price increase resulted in far less construction work and a considerable increase in the productivity of the work undertaken thereafter. The Building Cost Information Service tender-price index currently shows a reduction of approximately 30% in real terms in the price of new buildings over £40 000 since 1980, whilst the relevant cost indices have continued to increase. Thus long-term building contracts let at the time with a price escalation clause would currently cost the owners much more, if the work were tendered today. Owners considering price-variation clauses for long-term contracts will need to give this phenomenon considerable thought.

6.6 Conclusion

The current construction scene is noticeably different from that of a few years ago. More projects are being completed within budget and ahead of programme; there are relatively few new major projects. Competition is intense. Contractors are willing to accept greater risks, especially where they lie within their control.

This is illustrated by the willingness of UK civil contractors to accept the lane rental system described earlier, and by the major US architect engineers who promoted and would until recently only work in a cost-reimbursable system.

Employers in the USA are now able to obtain lump-sum quotations for multi-billion dollar projects from architect engineers acting as contractors — provided the project is not a nuclear power station!

This illustrates the need to consider the wide range of options available to an owner at the inception of a project, and the need to take into account possible changes in the market forces during the project period.

This Chapter seeks to encourage careful consideration of all the factors before commitment is made.

6.7 References

1 WEARN, S. H.: 'A review of reports of failures', *Proc. I. Mech. E.*, 1979, **193**, pp. 152–136 and suppl. vol. pp. S29–44

2 'Nuclear Construction: A Status Report', *New York Times*, 26 Feb. 1964

3 WEARNE, S. H. and NINOS, G. .: 'Responsibilities for project control during construction.' Report TMR 17, Technological Management, University of Bradford, 1984

4 PERRY, J. G. and HAYES, R. W.: Risk and its management in construction projects, *Proc. ICE*, 1985, **1**, pp. 24–28

5 BARNES, N. M. L.: 'How to allocate risks in construction contracts', *Int. J. Project Management*, 1983, **1**, pp. 24–28

6 'Large industrial sites' (National Economic Development Office, 1970)

7 'Guidelines for the management of major projects in the process industries', (National Economic Development Office, 1982)

8 'Fast track project management', *Int. J. Project Management*, 1984, **2**, pp. 240–241

9 HORGAN, M. O.: 'Competitive tendering for engineering contracts', (E & F Spon, 1984)

10 BARNES, M.: 'Lane rental', *Contract J.* 31 Oct. 1985

11 'Report of the Enquiry into the Pricing of Ministry of Aviation Contracts', July 1964

12 PERRY, J. G. and THOMPSON, P. A.: 'Requirments for target and cost-reimbursable construction contracts', Report 85, Construction Industry Research & Information Association, 1980

6.8 Appendix: Alternatives available — an outline contract spectrum

Type of contract	Advantage	Requirements	Problems	Employer's contract; administration resources
1 Lump sum	Employer knows the extent of his liability before he proceeds — thus whether his project is within his budget. Cost control delegated to contractor	The employer should know his requirements exactly	To the extent that requirements are not fully defined or when they are subject to variation, the employer puts the contractor at risk, and thus probably himself	Minimum
2 Part lump sum, part re-measure	Permits lump-sum prices to be obtained for substantial areas of precisely defined work, but offers the opportunity of assessing during the contract the cost of those areas of work which it was not possible to specify precisely at the outset	In addition to the lump-sum requirements, a schedule is included either for the pricing or variations and/or measurement	See 3(a)	See 3(a)
3 Re-measure	Permits the appointment of the contractor before full details of the design or ground conditions are known	An outline design indicating the extent of the works, but both parties understand that the contractor will be paid by measuring the actual extent of work; it is also understood that the works will not be varied beyond a reasonable extent		

6.8 Appendix: continued

Type of contract	Advantage	Requirements	Problems	Employer's contract; administration resources
3(a) Bill of quantities		Estimated quantities of each item of work extracted from the drawings. Provision made under 'preliminaries' for contractor's fixed costs and time-related costs; the contractor gives a rate for each item, and the extended total gives price for tender assessment	The contractor may price the bill to suit his expected commercial advantage rather than to reflect real costs, e.g. by: (i) including preliminaries in the rates (ii) 'loading' the price of early work (iii) increasing the rate for work judged to be underestimated. Difficulties arise if design or access is delayed, or if the nature of the work differs from contract drawings	Since, under this form of contract, the employer accepts responsibility for design, he requires architects and/or engineers; quantity surveyors are necessary to measure the work and probably to agree revised rates. It is usual to supervise work on site
3(b) Schedule of rates		Similar to above but quantities omitted; i.e. outline drawings indicating extent of work are accompanied by a price schedule. In an attempt to avoid distortion of pricing, some employers price the schedules from their own experience and invite tenderers to insert preliminaries and state the adjustment to sections of the schedule they require for the particular job	Similar to above but may be easier to administer if contract properly established	As above — possibly slightly less site administration, but requires the addition of a costing service to maintain the schedules

6.8 Appendix: continued

Type of contract	Advantage	Requirements	Problems	Employer's contract; administration resources
4 Fixed fee	Allows early appointment of contractor when extent of work is not sufficiently well established to permit a re-measure contract, or when the risks to the contractor are so great that an unacceptably large contingency would be included. Also used where a high degree of confidence has been established between the employer and the contractor, and where the removal of most risks from the contractor is thought to be to the employer's advantage. The contractor is encouraged to complete early and economically because his fee will not increase if additional time or resources are used	An outline of the project sufficient for the contractor to estimate the managerial resources and profit sought. The contractor is then paid his costs, plus the fixed fee which should not vary	(i) The employer will probably feel it prudent to compare costs with the budget, but is likely to feel inhibited if there are unexplained differences. (ii) If the job is varied or delayed for reasons outside the contractor's control by mor than a reasonable extent, the contractor may well feel entitled to a revised fee.	Design work and technical supervision generally as for measure and value; The cost has to be defined and monitored. Measurement not necessary except for cost effectiveness or productivity viewpoints

6.8 Appendix: continued

Type of contract	Advantage	Requirements	Problems	Employer's contract; administration resources
5 Target cost	An alternative to fixed-fee contracts, which seeks to provide a continuing incentive over a fairly wide band of work	A broad outline of the project. rates for the type of work involved are established in either bills-of-quantities or schedule-of-rae form; work is measured to establish a target price. A method of establishing actual cost is determined, and the way in which the difference between the target price and the identified cost (whether profit or loss) is specified	It must be recognised that the rates used for establishing the target price are likely to be only approximate. If this were not so, alternative form of contract would have been selected. The rates will tend to be high; if this is not recognised, the negotiation of rates may be time consuming and much of the advantage sought lost. Similarly, sharing of the difference between target and cost must spread over the likely difference. Requires considerable administration	In addition to engineering and site resources, requires both quantity surveyors and a means of assessing costs
6 Cost plus	Allows appointment of contractor at the earliest possible stage, and enables the use of contractor's services in planning the work	Minimum	Cost control	See 4
7 Day works	Permits competitive tender for labour, plant and material without defining work	Minimum	Work must be well directed by employer Cost control	Substantial

Chapter 7

Turnkey versus multi-contract

D. C. E. Brewerton
Kennedy & Donkin Group

7.1 Introduction

There are many bewildering choices facing the purchaser of a major project. This Chapter attemps to outline the two most common strategies currently being adopted for the implementation of large installations both overseas and in the United Kingdom. It discusses advantages and disadvantages of both methods from a project-management viewpoint, keeping in mind the three fundamental objectives of time, cost and quality. It is based upon the experience of a firm of independent consulting engineers who have been responsible for advising purchasers in their choice of the appropriate contract strategy and in assisting them to ensure a successful project conclusion. Although the Chapter is based largely upon experience in power generation, transmission and distribution, many of the conclusions are relevant to other industrial projects. it also provides an opportunity to set down some of the project management lessons learned.

7.2 The purchaser — aims and objectives

There is an enormous range in the levels of expertise of purchasers of plant and equipment throughout the world today. In the industrialised countries the purchaser may have his own project-management procedures well established with detailed instructions for the procurement of plant. The major electrical utilities usually have well developed methods for contracting and commissioning entire projects, whilst in other areas, particularly in developing countries, purchasers may have limited or no experience of modern project management.

The consulting engineer is frequently called upon to give advice at the very earliest stages of a project when it is no more than an awareness of need. In developing countries the finance is often provided by an international funding agency who will assist the purchaser in assessing the consulting engineer's proposals for carrying out feasibility or conceptual studies.

It is not unusual for purchasers to have difficulty in defining their needs, aims and objectives, and it is in this area that the consulting engineer is well placed to crystallise the requirements. In addition to defining the scope of the project three essential project management targets at the project outset must be established namely:

(*a*) *Time*: an overall time scale for realistic completion giving project milestones for stages of the work

(*b*) *Cost*: an estimate of the anticipated final cost and proposals for financing the project

(*c*) *Quality*: an assessment of quality requirements based upon the life and duty of the project

During these early stages of the project's life decisions have to be made upon the most suitable contract strategy. Although there are a large number of options to be considered, possibly the most important decision is whether to proceed on the bases of a single turnkey contract or to adopt a multi-contract approach.[1]

7.3 Turnkey — how it operates

The word 'turnkey' originated from the concept that a contractor would undertake to design and commission a complete installation until the time came for the purchaser to 'turn the key' for full operation.

In practice, the purchaser will have considerably more involvement than this simplistic definition suggests. However, the concept remains the same in that the contractor, in return for payment, undertakes all things necessary for the design and construction of the project, from inception to completion, ready for the use of the purchaser.

It can be seen immediately that there is one overriding advantage in this approach. Once the contract has been signed, the prime responsibility for the project rests with the contractor. A good turnkey contractor will produce a satisfactory project, but an inexperienced or poor turnkey contractor could prove to be a disaster for the purchaser and to himself.

To ensure that the best possible turnkey contractor is appointed, considerable care must be taken in his selection. Purchasers usually, therefore, engage an experienced consulting engineer to assist in prequalifying tenderers, in the preparation of enquiry documents and in the assessment of the tenders obtained. Prequalification procedures ensure that firms who are inexperienced or have inadequate resources are not invited to submit tenders. The enquiry documents provide an outline specification giving performance criteria for the plant and the quality of materials to be used for construction. Owing to the amount of work involved in preparation of turnkey tenders adequate time has to be allowed to enable contractors and consortia to complete sufficient basic design upon which to cost the project.[2]

During the tender assessment period the contractor's proposals for carrying out the work must be scrutinised very thoroughly. Once the contract has been placed the responsibility for design and construction becomes the contractor's. He can use his own methods for carrying out the work including any special expertise gained on previous projects. Usually significant elements will be subcontracted and the contractor's nominations for these subcontracts form an important part of the tender assessment. After contract award and during the detail design period, the purchaser's consultant is usually retained to approve the contractor's proposals to ensure that the design is fully in accordance with the specification.

Once the work has started on site, the purchaser will normally provide a representative from his own organisation, supported by a team of engineers from his consultant, to witness and check the quality of the work. This does not relieve the contractor of his responsibility for project management and all the interface engineering between the Contractor, and his subcontractors and suppliers.

7.4 Multi-contract — how it operates

A project demanding more than one contract can be termed 'multi-contract', but the expression is usually reserved for the larger projects where a number of contracts are involved. The concept is generally adopted because purchasers feel that they would prefer to have more direct control over the project and its design, including the selection of plant. Each component is subject to a separate contract which can be specified in some detail and selected in competition with other similar products. Many purchasers believe that, because of the competitive tenders obtained for each component, multi-contract offers a more economic project. Also, by influencing the tender list for particular elements, multi-contracts can more easily satisfy the political need for involvement of local contractors. Sometimes projects are multi-contract because a number of lending authorities are involved in funding it, each financing a particular part. Very careful planning of the various contracts is essential from the outset as the purchaser/consultant team assumes responsibility for the co-ordination between the principal contractors. The 'knock-on' effect of delay by one contractor to those following him can assume great significance, and it can quickly give rise to claim situations. A shorter tender period than for turnkey contracts is generally acceptable for each individual contract, but the overall period from inception to award of the last contract will be greater, as the contracts are usually placed progressively.

It is usual for the purchaser/consulting team to assume the responsibility for the detailed design of the civil-engineering works and the preparation of all working drawings. The timing of this essential design phase can be extremely critical to the project's ultimate success. It is common practice to award the main plant contracts early in the programme and to provide financial incentives or penalties to ensure that the civil-design interface information is available at the earliest possible date. Contracts for site clearance and bulk excavation can usually be arranged well in advance of the receipt of plant information.

The purchaser/consultant team will be closely involved in all aspects of the management of the project. They need to have available all the skill for controlling the various contracts, supervising the design work, checking the manufacture of the plant and its transportation to the project site, and supervising the installation and commissioning. Good opportunities are available to provide training for some of the purchaser's engineers by working alongside the consultant during the various stages of the project.

Experience has shown that the greater the number of contracts, the greater the risk of delay. This has to be balanced against value for money for individual components of the project. An optimum has to be achieved, taking account of the number of interfaces involved. Whilst it seems likely that for most projects about five or six 'island' contracts would be ideal, in practice there may be many more.

7.5 Tendering procedures

For a turnkey project, the purchaser/consultant team prepares a performance and quality specification covering the project through its various stages. For the multi-contract project the purchaser/consultant team decides the number and content of each contract, after completion of a basic design study. Factors that influence the decision on the number of contracts include the size and establishment of the purchaser's organisation, the finance arrangements to be employed, the extent of the purchaser's wish to influence the detailed design and plant selection, considerations of local manufacture and the overall programme time available. For both types of project the consultant must prepare a master programme, but in the case of the multi-contract project, each principal contractor will be required to schedule his work in accordance with the master milestone time schedule.

Although there are model forms of conditions of contract for electrical and mechanical works and separate conditions for civil engineering works, there are no such generally recognised conditions for turnkey projects.[3] These usually have to be prepared by the purchaser or his consultant based on past experience and the current project needs. It is essential to make very clear the design responsibilities of the contractor as well as his project-management responsibilities. Pre-qualification is usually carried out to keep the list of bidders to sensible limits. At present, there are only a limited number of suitable firms in the power industry capable of undertaking major turnkey projects.

Guidance is given in Table 7.1 for minimum periods to prepare, price and assess tenders. In some cases the longer periods shown in the Table may not be available owing to severe time constraints. These are frequently caused by irretrievable delays at project outset, in obtaining the necessary funding or arising from incorrect decisions on contract strategy. Therefore, on major overseas projects a good turnkey contractor will obtain market intelligence before receipt of the formal enquiry to enable him as a commercial risk to commence preparation of his offer in advance.

Table 1   *Minimum time periods for tendering and tender assessment*

Contract type	Weeks for		
	Preparing inquiry	contractor to produce tender	tender assessment
Multi-contract			
Mechanical and electrical plant contract with system design	8–12	10–14	6–10
Civil-engineering contract with full bills of quantities	8–10	8–10	4–6
Turnkey			
Large multi-discipline	8–10	12–18	12

Even longer periods may be required for preparing tenders for large turnkey projects such as major petrochemical plants and urban metro schemes, but for most power stations and electrical projects the periods given are adequate. Once the tenders have been received the purchaser/consultant team carries out a detailed adjudication of the bids received for the work. The analysis must be very thorough to enable a proper comparison to be made. It is common practice to invite short-listed tenderers to separate meetings after carrying out an initial assessment of the tenders submitted, to clarify any points in their bid likely to affect the final recommendation. Often tenderers arrive with large numbers of engineers at such meetings to ensure that all questions are properly and swiftly answered.[4] The purchaser/consultant team must, of course, be prepared or able to match the contractors' expertise at this critical stage. Procedures are required to ensure all tenderers are asked the same questions and not given the opportunity to vary their offer commercially in an attempt to gain unfair advantage over their competitors.

All factors have to be taken into account when assessing tenders, but in the case of turnkey contracts special financing terms offered by tenderers often become particularly significant. Examples of this are the soft-loan facilities offered by Japanese contractors and the government aid programmes favouring some European countries.[5] Attractive terms are sometimes offered for the larger 'island' contracts in the multi-contract situation, especially where there is an opportunity for exporting major elements of plant.[6]

The time taken to assess the tenders and to complete the negotiations leading up to contract award can vary greatly. Computer software is available to assist in assessing the often complex financing terms, and as a broad guide Table 7.1 gives an indication of the minimum time needed.

7.6 Programme after contract award

Once the contract has been placed it is likely that a turnkey project will be completed in a shorter time than a multi-contract project. The major reason for this is that a turnkey contractor, having less restraint on sub-contractor selection, can make such appointments without calling for international bids and making formal comparisons for consideration by the purchaser. The turnkey contractor can also save time on the civil design because he can short-circuit the formal procedure of passing plant information to the purchaser/consultant team. Being a producer of prime civil data, he can pass this information in piecemeal fashion to his chosen civil partner or sub-contractor. This team application can also assist in speeding up works in the latter stages of the job, and particularly during the erection and commissioning phases.

On a multi-contract project, the collection of information by the purchaser/ consultant team for analysis and for passing on to other contractors is a significant factor in the overall programme of activities. A plant contractor has more incentive to give priority to long-delivery plant item design to safeguard his own position, rather than prepare civil design data for others. However, this problem can be overcome by establishing 'design freeze' dates for the exchange of essential interface data and linking these to penalty or, preferably, to bonus payment clauses in the contract.

A major feature of the programme for each of the contractors on a multi-contract project is the requirment for providing clear access onsite for the following contractor's construction and erection activities. Experience has shown that, under the multi-contract system, individual contractors display extreme reluctance to carry out work in areas in which other contractors are already working. The purchaser/consultant team must manage the project to ensure that inefficient working arrangements and subsequent claims for waiting time do not occur.

Despite the potential programming difficulties with multi-contract, the purchaser/consultant team is able to maintain more effective control over the scope, content and component parts of the project, and is more likely to create an end product closer to the purchaser's own particular requirements.

Table 7.2 sets out in bar-chart form typical overall programmes for the two types of contract.

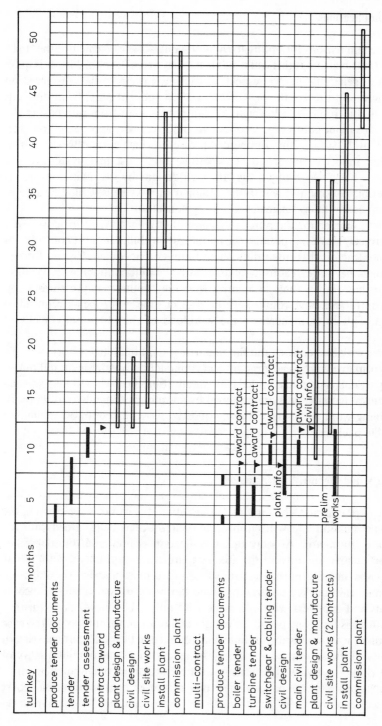

Table 7.2 *Simplified turnkey and multi-contract programmes for power station (one unit only)* Note that periods are indicative only and shorter times may be achieved in the future

7.7 Cost implications

One of the main reasons for opting for a multi-contract arrangement rather than a turnkey is the belief that this will result in a lower overall capital cost. It is quite erroneous to generalise in this respect, as each case needs to be considered on its own merits. In theory, the procurement of component parts from different suppliers should result in a lower total capital cost than from purchasing the total plant from one contractor. Proponents of the turnkey approach, however, maintain that, because of the commercial pressures which currently exist world-wide within industry, an equally competitive total cost may result from a turnkey bid. They also contend that the civil designs by the main contractor, kept under his control and within the commercial risks he is willing to accept, may be less costly than designs executed by consultants.

It is much simpler for the purchaser/consultant team to monitor costs on a turnkey project. However, because the major responsibility for cost control rests with the contractor, it is often difficult to obtain an adequate degree of detail which would be more readily available with a multi-contract arrangement. Experience has shown that, in many cases, the commercial pressures upon turnkey contractors have resulted in the project supervision being reduced to save costs. This in turn increases the workload of the purchaser/consultant team.

Problems with cost control and finance arise if it is attempted to form a single contract turnkey for the very large project running into hundreds of millions of pounds, particularly if a fixed price is required. It has been shown that, in such circumstances, better results are obtained by adopting an island-contract basis, thus enabling work to be started more quickly by letting contracts in smaller manageable packages.[7]

7.8 Quality and technical performance

With a turnkey contract, the standard of quality of the works is set down in the original specification and written into the contract document agreed with the appropriate contractor. As only one specification is prepared, the degree of detail, particularly for some of the auxiliary equiment, is usually less than that which may be included in the relevant section of a multi-contract specification. This permits the turnkey contractor some flexibility in the overall choice of equipment and suppliers with whom he is familiar. Therefore, on a turnkey contract there is less control on the overall choice of suppliers than is the case on a multi-contract arrangement, but if care is taken to examine the track record and quality of the sub-suppliers during the tender stage, the overall quality of the plant should not vary significantly with either contractual arrangement.

For most projects involving a high proportion of civil-engineering work, the choice, capability and performance of the responsible civil-engineering sub-contractor will be very significant. Once a turnkey contract is placed, the purchaser/consultant team may not be able to influence the selection of the civil sub-contractor; his capabilities must therefore be thoroughly explored at the tender stage. Quality control during the currency of a turnkey contract will be obtained by formal submission of designs to the consultant for approval, together with details of the materials proposed for use in the works. This, together with site surveillance by the purchaser/consultant team, will ensure that the specified quality of works is maintained. However, the technical quality of the civil works for a multi-contract will be directly under the control of the consultant, who will prepare the design and detailed working drawings.

7.9 Duties and responsibilities of the purchaser

This is one of the most important areas for consideration when deciding whether a project should be engineered on a turnkey or multi-contract basis. Whilst it would be possible for a small recently established purchasing organisation to handle the work involved on a turnkey project, it would be difficult for an inexperienced or small organisation to handle a multi-contract project with the inherent complexities of, for example, a power station unless they engaged a consultant to assist them. Having concluded the formal arrangement for a single turnkey contract, the purchaser may take as little or as much further interest as he may thereafter desire or have the capacity for. With the multi-contract package arrangement, the purchaser cannot adopt an inactive role. By virtue of the number of packages, he is involved in the issuing of more enquiries, the review of more bids and the placing of more contracts. Further, he is required to provide basic and interface data to enable each of the contractors to work within specific limits of responsibility. Another of the purchaser's tasks is to provide the initial site layout and the information on essential services which may be provided by, say, the civil contractor to each of the subsequent plant contractors. In addition, the purchaser will need to review at the various stages of the project the inevitable engineering variations which arise in ensuring that the various packages neatly fit together to form a composite unit. It is difficult when ordering a number of packages in parallel to ensure that each is totally compatible with the other prior to the placing of the order. On the multi-contract arrangement, the purchaser must also accept some of the risk for the overall project, which in the case of a turnkey contract is taken by the contractor. For a multi-contract project it is impossible to escape the conclusion that the design of the civil-engineering works and buildings, and the preparation of the working drawings, must be the purchaser's responsibility.

In practice, the purchaser who decides to proceed on the multi-contract basis, but who does not have the resources to carry out these tasks, will use his consultant to do them for him. The purchaser's tasks will then be limited to such matters as the issue of contract-award letters (drafted by the consultant) and making decisions on operating procedures. He will also be responsible for making payments to the contractors, which may involve different terms from one contract to another. This could involve the purchaser in having a reasonably sized accounts or contracts department. In summary, the organisational involvement of the purchaser needs greater composition and stability for a multi-contract project than for a turnkey arrangement.

7.10 Duties and responsibilites of the consultant

As for the purchaser, the duties and responsibilities of the consultant, where he is appointed, vary significantly for the turnkey and multi-contract approaches. There are differences in the consultant's responsibilities during the pre-contract stage. In the multi-contract case these mainly consist of the additional work involved in the preparation of the specification documents and the additional activities in ensuring that co-ordinated and compatible bids are obtained. Similarly, additional monitoring activities are required during the post-contract stage because of the larger number of contractors involved.

On the multi-contract arrangement, considerable attention needs to be paid by the consultant to the interfacing between the respective packages, and close planning is required for all the co-ordinated activities. The consultant probably also has to provide the civil-engineering designs and working drawings. The additional project-management efforts places considerable onus on the consultant, who must be given sufficient time and facility to execute this responsibility if the job is to be ultimately successful. All of this work comes over and above the normal surveillance role which is exercised on a turnkey contract.

The co-ordinating responsibility extends through the phases of design, manufacture, construction, erection, commissioning and testing, and therefore, results in the consulting engineer needing to provide more resources for the multi-contract arrangement than for the turnkey. In essence, the turnkey contractor takes over much of the responsibility which would have been provided by the consulting engineer for complete project management and for ensuring that the job both fits together and is completed on time.

A completely new role for consulting engineers has developed in recent years as a direct result of the increased popularity of turnkey projects. This is the provision by consultants of design and other services to supplement the skills of the contractor and to enhance his bid. A consultant may be retained throughout all the stages of the project by the turnkey contractor, working alongside him during the design period and sometimes providing site staff and commissioning engineers. Consultants providing these services to contractors have to ensure that there is no conflict of interest with any agreements that may exist with purchasers on other projects. Provided these conditions are satisfied, purchasers generally view favourably turnkey tenders that include the technical expertise of a consultant.

7.11 Conclusions

It can be concluded that a number of factors need to be considered before deciding whether to proceed on a turnkey or multi-contract basis. Using the three prime objectives of project management, the principal factors can be summarised as follows:

7.11.1 Time

More time is required for tenderers to prepare bids for turnkey projects, and more time is needed to assess the bids and place a contract. Once the contract is placed, the turnkey project should proceed more rapidly than the multi-contract. This is particularly the case if a proven design is being offered. However, good results in terms of progress can be obtained with the multi-contract approach in developed and developing countries, provided the purchaser/consultant can establish a substantial project-management team capable of making the decisions that are required swiftly and confidently.

7.11.2 Costs

Although apparent savings in cost may be made by using the multi-contract method, current commercial pressures worldwide can produce an equally competitive total cost using the turnkey approach. In practice, a multi-contract may be more costly because the purchaser would choose the best quality and limit tenders to select lists of suppliers. Funding-agency assistance is more readily available in certain countries for turnkey projects than for small packages on multi-contract work. Finance houses and government aid provisions have, in the past few years, been most attracted to total job concepts where the contractual responsibility lay with a single organisation.

However, for the very large project involving bi-lateral aid, the work sometimes has to be split into island contracts, and this results in a multi-contract or a limited turnkey approach.

7.11.3 Quality

More direct control of quality is possible with multi-contract projects, as the purchaser is in a stronger position to influence the scope, content and component parts of the projects. However, the purchaser can obtain adequate control on a turnkey project by engaging an experienced consultant to monitor and approve the work throughout the many stages, from feasibility study through to final commissioning and testing. It is essential to define the role of the consultant in the terms of the turnkey contract. The method lends itself well to the concept of 'quality management' outlined in BS 5750, which requires the contractor to submit to the purchaser details of his quality plan for approval, and to provide systems and procedures for maintaining quality throughout the life of the project.

In view of the advantages in the turnkey approach, particularly in developing countries overseas, the absence of any suitable internationally recognised and acceptable conditions of contract is likely to give problems in the future. Only the widely experienced purchasers and consultants are likely to have developed suitable conditions for use on turnkey projects which encompass the particular needs of the purchaser, the scope of the work to be carried out and the rules of any funding agency which may be providing assistance under a loan agreement. Because of the heavy burden on contractors faced with the task of preparing turnkey bids and the large tendering costs involved, it is essential to standardise as far as possible the tendering procedure and to ensure a common basis for preparing and comparing bids and for improving the management of turnkey projects.

The final decision on turnkey versus multi-contract very often depends upon the requirements of the funding agencies, but when it is possible to make a choice, a careful assessment of the risks involved must be made. Generally, a turnkey project is faster than a multi-contract project, and is unlikely to be more costly owing to the intense competition that currently exists worldwide and pressure from suppliers anxious to keep their production lines and order books full. Most of the risks can be eliminated by rigorous assessment at the tender stage. The extra time and cost of the multi-contract is compensated by better quality through greater freedom of choice and greater powers of the purchaser to insist on the best.

7.12 References

1 MARSH, P. D. V.: 'Contracting for engineering and construction projects' (Gower Press, 2nd edn. 1981)
2 'Tendering procedures for evaluating tenders for civil-engineering projects'. Federation Internationale des Ingenieurs-Conseils, Lausanne, 1982. Also HORGAN, M. O'C.: 'Competitive tendering for engineering contracts', (Spon, 1st edn., 1984)
3 The nearest examples to consider using are the Model Conditions of Contract for lump sum complete process plants in the UK published by Institution of Chemical Engineers (revised version, 1981). Model forms for turnkey contracts for complete fertiliser plants have been published by UNIDO, 1982
4 During Dubai E power station tender assessment for the Dubai Electricity Company carried out by Kennedy & Donkin in January 1985, both Japanese and Korean tenderers provided up to 40 people to discuss their turnkey bids
5 Soft loans were arranged by the Japanese for the turnkey Batangas power station in the Philippines, and Italian aid provided for phase 2 of the turnkey Ras Katenib power station in the Yemen Arab Republic (consulting engineers: Kennedy & Donkin)
6 British aid was provided for the turbine 'island contract' and German/Scandinavian aid for the boilers for phase 1 of the multi-contract Morupule power station in Botswana. For phase 2 the turbine contract attracted Japanese aid in the form of a soft loan and Scandinavian aid was provided for the boilers (consulting engineers: Kennedy & Donkin)
7 A good example of this is the Hong Kong Mass Transit Project. The Japanese turnkey contractor withdrew his original fixed-price bid and the project was re-let on a multi-contract basis (consulting engineers: Kennedy Henderson Limited for mechanical and electrical; in association with Freeman Fox & Partners)

Chapter 8

Quality assurance and project management

R. M. Macmillan
Central Electricity Generating Board

8.1 Introduction

The CEGB capital expenditure for the year ending 31 March 1985 was £887 million, just over half of which was on nuclear power plant. Strict financial practices control the spending of such large sums, which, for new power-station or transmission plant or for significant projects associated with existing plant, require the monination of CEGB responsible officers. Each officer is formally appointed, and his principal responsibility is to ensure the execution of his project and be answerable for completion at the due date, within cost and with the specified quality and performance. He also reports on progress and experience likely to influence future policy.

8.2 Policy

It is the policy of the Central Electricity Generating Board that for all items of power generating and transmission plant and associated systems there shall be in force appropriate arrangements for providing assurance of quality at all stages from design to decommissioning.

This means that the CEGB, and those who provide its plant, equipment and services, must develop and apply sound quality-assurance practices, and that decisions must be taken on the actions necessary to provide confidence that the plant and equipment will be safe and perform satisfactorily throughout its life.

For nuclear-safety-related plant it is the policy of the CEGB to comply with BS 5882 'Specification for a total quality assurance programme for nuclear installations', and with minor exceptions to require contractors supplying such plant also to meet the requirements of the standard. BS 5882 specifies quality-assurance requirements throughout the lifetime of the plant, and which are therefore applicable to the plant owner and his contractors. For other plant, the CEGB still bases its practices on the quality-assurance factors identified in BS 5882, and requires contractors to comply with the appropriate part of BS 5750 'Quality systems'. Thus for each power station or for each significant project, nuclear or conventional, the CEGB develops and applies a quality-assurance programme.

8.3 Background

Since the formation in the early post-war years the CEGB have taken steps to satisfy themselves that plant and equipment of the required quality was being delivered by manufacturers to CEGB sites. This typically involved detailed inspection and surveillance at manufacturers' works.

During the 1960s, when a considerable number of large projects were under construction, it became apparent that inspection, although of utmost importance, was not in itself sufficient to provide adequate assurance that plant and equipment would be satisfactory in service. In addition to the inspection of plant, therefore, increasing attention was paid to the adequacy of the manufacturer's management controls, or, as it is now termed, his quality-assurance system. Assistance was given to manufacturers in the development of appropriate systems, and achievement of required quality resulted. These were particularly relevant to process controls such as welding, and there was a significant reduction in failures attributable to unsound welds and poor or rogue material. The use of these disciplines was still centred on 'guidance' and 'recommendations' for the assessment of potential suppliers, supported by inspection and test arrangements, but experience with, for example, Hartlepool, Heysham 1 and Grain power stations demonstrated that the Board needed to adopt a much stronger contractual approach and to develop and specify firm 'requirements'.

In 1979 BSI issued BS 5750 — 'Quality Systems', which has subsequently formed the cornerstone of the Government's Quality Campaign. Further related developments have been the DTI Register of Assessed Suppliers and the National Accreditation Council. These latter initiatives have been paralleled by the International Organisation for Standardisation (ISO), who are drafting documents which are compatible with BS 5750 and which will assist UK manufacturers to tender for contracts in overseas markets.

BS 5882 is based on ISO 6215 'Nuclear power plants: quality assurance' and is compatible with the International Atomic Energy Authority (IAEA) Code of Practice and its associated Safety Guides. BS 5882 meets the Nuclear Installations Inspectorate (NII) Guide to Quality Assurance Programmes for Nuclear Installations and has been written into the Sizewell B Pre-construction Safety Report Quality Assurance Arrangements.

8.4 Application to construction projects

The management of a project includes two important quality objectives Firstly to define, in technical specifications, plans and manufacturing procedures, the required quality of the plant, and secondly to ensure that this required quality is obtained from the contractors responsible. Quality-assurance principles require that throughout the project the achievement of quality is continuously demonstrated and verified.

During the design phase of the project the responsible officer must ensure that the management organisation and the activities associated with the design of the plant are described in a quality-assurance programme. This programme defines the organisational structure, line and functional responsibilities, levels of authority, lines of internal and external communication and the necessary qualifications of personnel achieved through their education, training and experience. Arrangements are described for the review of all aspects of the design proposals, including design modifications and requirements for inspection during plant operation.

The responsible officer must be satisfied that design contracts will result in designs of plant and equipment of the required quality, and also that designs developed with the CEGB result in technical specifications that identify the required quality. There must therefore be in place arrangements which, in addition to the independent verification of critical calculations, provide assurance that requirement such as codes, standards, functional safety, reliability, manufacturing and inspection are correctly translated into specifications (i.e. drawings, procedures, acceptance criteria, test instructions etc.). Designs are reviewed to ensure that they are clearly capable of manufacture using the existing plant of the supplier or with clearly identified new plant using proven techniques. It is further necessary to ensure that these designs are capable of inspection during manufacture and operation using approved methods and equipment, and that they are capable of easy maintenance with readily obtainable replacement parts. Where some or all of these requirements have not been met, an adequate development programme must be prepared.

During the supply phase of the project the responsible officer must evaluate the quality-assurance capability of tenderers, ensure that the quality-assurance effort required of the contractor is clearly understood by him and is achievable, and he must agree with the CEGB inspector the manner and extent of verification required during manufacture and construction. The quality-assurance programme covering the supply of plant describes or identifies arrangements for compiling tender lists and assessing the tenderers' capability, including, for critical plant areas, the capability of proposed sub-contractors.

Tenderers may already have had their quality-assurance systems evaluated within the CEGB Certificate of Compliance Scheme (which covers the Board's main contractors, principal sub-contractors and suppliers of strategic materials or components), and proposed sub-contractors may have their quality-assurance systems certificated by one of the various national or sector schemes. This knowledge is considered, together with any relevant contractor performance data, and, where judged necessary, a visit is made to a tenderer to assess the current effectiveness of his systems. Inquiry specification clauses require tenderers to submit for assessment their quality-assurance programme (to meet BS 5882 requirements for nuclear safety related plant and equipment) or arrangements (to meet BS 5750 requirements for all other plant and equipment). Proposed sub-contractors must be listed, together with the basis on which the adequacy of their quality-assurance arrangements will be ensured by the tenderer. The inquiry specifications lead to contract commitments which also require the contractor to submit to the responsible officer, for his approval, quality plans which identify manufacturing operations, the controlling procedures, the inspections and tests which will be carried out and the records which will be raised.

The role of the CEGB inspector is a most important one. It is he who will, on behalf of the responsible officer, verify that specified quality is being achieved during manufacture and construction. In addition, he provides valuable information on the tenderer's ability to meet the technical specifications and the quality-assurance requirements. The CEGB approach to quality assurance during the supply phase of the project is based on a close relationship between the responsible officer and the inspector.

On significant projects the inspector is involved at the initial planning stage, and he provides comment to the responsible officer on tender lists, tender submissions, the technical specifications, quality-plan requirements, manufacturing and testing procedures, sub-contracting proposals and documentation requirements. In addition, using his knowledge and experience of the proposed contractor, he informs the responsible officer of the monitoring he proposes to undertake during the course of the contract. These proposals are discussed and agreed, and are marked up on contractor's quality plans. The quality plan is therefore an important controlling document within the CEGB quality-assurance approach. The dialogue described takes place on all projects/contracts judged by the relevant responsible officer to be 'significant', which typically means any area where there could be safety or plant-availability implications or where major expenditure is involved.

The discussions between the responsible-officer/engineer and the inspector come under the generic title of 'bipartite meetings', and another important concept in applying sound quality-assurance practice to CEGB projects are 'tripartite meetings'. These bring together, often prior to contract placement, the responsible-officer/engineer, the inspector and the contractor (and perhaps major sub-contractors). The purpose of these discussions, which may continue at regular intervals during the execution of the contract, is to resolve any outstanding matters relating to the contractor's quality-assurance programme or arrangements, to establish the nature and extent of control to be exercised over sub-contractors, to make known to the contractor the verification activities to be carried out by the inspector at the contractor's and sub-contractors' works and on site, and to resolve any quality-related problems which may be evident.

The CEGB quality-assurance programme describes the above arrangements and is supported by the quality-assurance programmes submitted by contractors and by the inspection organisation.

8.5 Audit and review

Quality-assurance principles require that compliance to, and effectiveness of quality-assurance programmes and arrangements are verified by internal audit. External audit may also be required to be carried out by authorized organizations. The responsible officer therefore develops arrangements for the internal audit of his programmes, and the CEGB have trained and qualified staff to perform the task, who are listed in an internal register and they are also qualified against the Institute of Quality Assurance criteria. In addition, the Board has in place arrangements for auditing of project quality-assurance programmes by organisationally independent groups.

CEGB quality-assurance programmes are also subject to audit by external organizations; e.g. programmes for nuclear-safety-related plant are audited by HM Nuclear Installations Inspectorate (NII), who oversee licensing arrangements and the maintenance of licence requirements on behalf of the Health & Safety Executive. Similarly, contractors' quality-assurance programmes or arrangements are subject to internal audit control, and may be audited by the CEGB and in some instances the NII. Corrective actions are implemented and reviewed after each audit.

8.6 Plant operation

The CEGB approach to quality assurance recognises the importance of maintaining the required quality while plant is being commissioned, operated, repaired, refurbished and finally decommissioned. This is an aspect of quality-assurance application which in many industries has received little or no attention. Within the CEGB, quality-assurance programmes for the operational phase of plant life are currently being developed, both for Sizewell B, where it will be a licence condition, and for all our currently operating plants. Programmes for new stations recognise the need for continuity with project programmes, and describe arrangements for consultation with the responsible officer on matters relating to commissioning, operation and plant maintenance, and identify procedures for the hand-over of plant and associated documentation.

In many respects BS 5882 lacks detail coverage of plant operation, and the Board has found it necessary to expend considerable effort in developing documents which give guidance on appropriate quality-assurance practices in this important area. During the work reference was made to the IAEA Safety Guide 50-SG-QA5 'Quality assurance during operation of nuclear power plant', and it is expected that the experience gained will be taken into account during the forthcoming review of BS 5882. Operational quality-assurance programmes identify arrangements for the control of plant operation, maintenance, modifications, provision of spares, the letter of contracts and records, in addition to describing organisational responsibilities and personnel qualifications and training.

8.7 Application to Heysham 2

It is appropriate to describe the application of these principles to the latest CEGB power station now approaching completion.

At the commencement of the Heysham 2 project is was soon apparent that attitudes were present, throughout the organisations involved, which varied from strong resistance, to an apathetic acknowledgement of the existence and objectives of 'formal' quality assurance. Perceptions were that quality assurance costs time and money, generated paperwork, and diverted much needed effort and concentration away from the actual 'doing' of the work. Certainly in the early days, the pressures of re-generating a somewhat moribund industry and restoring morale, the inbuilt traditional resistance to 'outside' people measuring and possibly criticising individual or organisational performance, all combined to support this. Experience has shown that, even without the advent or imposition of quality-assurance practices, the organisation proposals necessary to attain the project objectives would have required the provision of comparable formal management procedures and documentation.

As the project has progressed, the increasing value of the close control and co-ordination of activities arising from the use of these procedures has become evident, and in particular the control of design innovation, evolution or change has made an important contribution to reducing the number of changes, and when changes are inevitable, ensuring that everyone who needs to know does know, understands the reasons and takes appropriate action. Previous experience demonstrates that this activity has not been very successfully managed or controlled, and that it has been responsible for serious defects on early projects.

The Heysham 2 project team implemented a system which not only controls design changes but also ensures financial and commercial compatibility. The procedures reinforce previous practices in the management of cost, programme and associated project activities, and, with a steady build up or change of resources, the availability of the system and associated documentation allowed early education and indoctrination to everyone involved, resulting in a more effective contribution.

One difficulty has been a reluctance to provide dedicated resource to progress and manage quality-assurance activities, such as handling responses to audits, quality plans and managing service agreements. Even with a committed resource, which has been implemented on the Heysham 2 project, the required contribution is sometimes given a priority which does not always achieve the ideal quality-assurance progress.

There is, however, a danger in over-emphasising the role of the dedicated resource or unit within an organisation such as a project team, in that it can deflect from what should be an essential attitude which every member of that unit should adopt; i.e. that quality assurance should not be considered as something quite separate from their traditional activities or services within the project — Implementation of quality assurance is the responsibility of every project team member. The members of any unit should be consciously aware of what is embodied in the quality-assurance programme and procedures, and so should be committed and active in ensuring compliance with quality-assurance requirements. The provision of a separate section within a project team would tend to dilute this attitude. It is for this reason that the CEGB divisional quality assurance unit confines itself, internal auditing apart, to giving functional advice.

Despite the positive commitment to quality assurance, CEGB experience, which is by no means unique, is that the resolution of audit findings is often an exceedingly slow and difficult process. It is difficult to be specific on the reasons for this, but generally they are a result of the hierarchical nature of the system and the lines of communication; for instance, between the NII and a nuclear island sub-contractor there are at least five interfaces. Even with dedicated resource to act and progress the necessary information within each organisation, a minimum period of 13 weeks is required for transmitting and replying to one piece of correspondence which crosses these boundaries.

Predictably perhaps, audits on this project have, without exception, revealed non-conformances with the agreed procedures. Also without exception, these revelations have proved valuable and made a positive contribution to the achievement of the project objectives. Audits provide positive proof of the effectiveness of the system and have resulted in much improved control of, and information from, the Board's agents and contractors. Initially results were often achieved by the auditor in the face of defensive or aggressive attitudes on the part of the auditee. However, this problem diminished progressively with the greater appreciation of the value of the system.

Experience has shown that programmes and procedures which appeared quite suitable at the commencement required re-writing or modifying, to conform to the on-going practical requirements of the project. It is usually as a result of audits, i.e. views, on actual practices compared with the pre-conceived requirements, which identifies that these are impractical, and hence need modifying. Where these documents are of HSE approved status in accordance with site licence conditions, revisions require formal procedures to be followed. Where they are not of approved status, changes can be agreed and implemented in accordance with internal procedures.

Recognising the changing nature of the situation and the need for continuing education, the practice on the Heysham 2 project is that quality-assurance performance is discussed at all project internal and interface meetings. Internally a selected procedure is discussed each month and any comments considered and modifications agreed. This ensures updating of practice and more important, an awareness of the procedure. Experience has shown it is essential that the people who actually use the quality-assurance programmes and procedures must have the prime input to the initial preparation and any ensuing modifications of these documents.

8.8 Conclusions

The supply, installation, setting to work and use of safe and reliable plant and equipment require that it be designed, supplied, constructed, commissioned, operated, maintained and finally decommissioned within the framework of a management system which defines responsibilities and ensures that the required quality of plant and equipment at each stage is properly defined, is obtained and maintained. The CEGB believe that the application of sound quality-assurance practices will significantly assist the achievement of this object. As with all organisations which have adopted this approach, we have been quick to realise that the simple words and sentences within the standards hide a complexity of interpretation and scope of application. Considerable effort has been, and is still being, applied to ensure that line management understand their responsibility for having in place appropriate arrangements for the assurance of quality. The Heysham 2 project has benefited considerably from the controls and disciplines via formal quality assurance. It is expected that the more comprehensive arrangements for Sizewell B will not only provide the CEGB with the assurance of a successful project, but will give assurance to those outside the CEGB that the safety and reliability requirements have been met.

Chapter 9

*Quality assurance

C. A. Mills

277 West 31st Street, Hamilton, Ont. Canada

9.1 Introduction

9.1.1 Background

The project-quality plan, described in this Chapter, was initially developed by the author as a means of defining in the specific quality progammes for major military contracts. The quality programme of the organisation was a total quality progamme[1] designed as a cost-efficient system to meet the requirements of AQAP 1.[2] Therefore, it was necessary to develop a comprehensive quality plan integrating into a single document all the differing controls and verifications called up by the procurement quality standards and contracts. These included such titles as inspection plan, test plan, design reviews, reliability programme plan, reliability demonstration, maintainablity demonstration, product verification etc.

This comprehensive document was developed as a replacement for the various individual plans in an endeavour to reduce the probability of errors of omission and duplication during the project design and production life cycle. Either of these types of errors can lead to significant delays and cost increases.

Since the initial use of military contracts, the approach has been successfully used on various high-technology commercial and industrial contracts and proposals. This approach has also been adopted by other organisations in meeting the requirements of the Canadian Quality Program Standards CAN3-Z299.1 and CAN3-Z299.[3]

9.1.2 Purpose

The project quality plan and the project definition or specification are complementary documents and *not* in conflict with each other. Both should be controlled documents, forming top-level items in the project data package; i.e.

---

* British readers should note that the project quality plan referred to in this Chapter is closely analogous to a contractor's quality-assurance programme under United Kingdom arrangements referred to in Chapter 8 by R. A. Matthews.

both should be subject to the same control procedures for origination, approval, distribution and change.

The project definition or specification defines, in quantitative, measurable terms, the requirements for the end item or service. These can include such facets as performance, operating requirements (architectural and human), availability, reliability, maintainability, working environment, shipping or storage environments, finishes, material restrictions, etc.

The project quality plan complements the above by customising the general quality programme[4] of the organisation to the needs of the particular project. It selects and defines the various control and verification activities necessary to give management and the customer confidence that the end item or system is capable of satisfying the needs of the customer; i.e. the desired quality has been achieved. The plan should *not* duplicate any of the acceptance criteria included in the project specification or other lower-level documents. Obviously, this is to prevent any errors due to double specification.

9.2 Definitions

To ensure a common understanding of the terms used in this Chapter the following definitions apply. The References are given so that readers are aware of the sources used:

(*a*) *Quality*: The totality of features and characteristics of a product, process or service that bear its ability to satisfy stated or implied needs.[5]

(*b*) *Quality system*: The organisational structure, responsibilities, procedures, activities, capabilities and resources that together aim to ensure products, processes or services will satisfy stated or implied needs.[5]

(*c*) *Quality programme*: The documented plans for implementing the quality system.[6]

These include:

(i) *Quality manual*: A publication defining the quality policy of the organisation and the contributions, responsibilities, acountabilities etc. of the various segments of the organisation under the quality system. The manual may include the procedures necessary to implement the intent of the manual. However, this is generally inadvisable as policies, responsibilities, accountabilities etc. are less prone to change than are controls or procedures. NB. A quality manual is a commitment by management and all personnel to develop techniques for doing the job right first time and providing the control and verification activities necessary to maintain and improve the project or process quality.

(ii) *Quality procedures*: Those procedures necessary to define the control and verification activities of the quality system. These provide the benchmarks for verifying the conformance to the system. They also form the basis for measuring the effectiveness of improved methods. Typically, these will include procedures relating to the marketing, design, material, production, installation or commissioning and quality, etc. functions, to the extent that these functions are involved in the intent of the business.

(iii) *Quality plan*: A document setting out the specific quality procedures, practices, resources and activities relevant to a particular product, process, service, contract or project.[5]

(*d*) *Quality assurance*: All those planned and systematic actions necessary to provide adequate confidence that a product, process or system will satisfy given quality requirements.[5]

(*e*) *Quality control*: The operational techniques and activities that are used to satisfy quality requirements.[5]

9.3 Why a project quality plan

The project quality plan provides a comprehensive integrated documentation of all the various control and verification procedures that apply to that specific project. In effect, it customises the more general requirements of the quality manual and procedures manual to the specific needs of the project.

On large projects, numerous controls and verifications are necessary to provide confidence that the resulting output will have the desired quality; i.e. it satisfies the needs of the customer. Some of these will clearly apply to certain segments of the organisation; others may not be so clearly defined.

The project quality plan should clearly define who is responsible for each particular activity and where in the project life cycle it should take place. This will remove any ambiguities or misunderstandings that may otherwise exist. This action should prevent, or at least reduce, the risk of errors due to the omission or duplication of controls or verifications. This reduction in risk should increase the probability that the desired quality can be achieved within the allowable cost and schedule limitations.

The use of the project quality plan as an integrated planning document may originate with the management of the organisation concerned. However, usually in the first instance it will be a development from a contractual requirement for verification plans in some form. Project quality plans by one name or another are requirements of most procurement quality system standards e.g. Canadian CAN3-Z299 series, ISO DIS 9000 series, NATO AQAP series, BSI 5750 series etc.

Contractual requirements may consist of a simple inspection plan for a contract with simple material and manufacturing activities. However, they may also consist of a series of plans such as would be required for a major system or operating establishment, like a power station, where plans covering reliability demonstration, reliability improvement or growth, maintability demonstration, availability demonstration, design reviews, product verification, etc. may also be required.

As one examines the total quality-assurance situation, it becomes apparent that numerous plans referenced outside the procurement quality system standard are, in fact, elements of the quality-assurance programme. Therefore, plans such as those involved with design reviews, reliability, availability and maintainability normally referenced in other documents should be included in the overall project quality plan. This will ensure these activities are efficiently integrated into a single programme.

In some cases it may take a few errors of omission or duplication, resulting in project delays and added costs, to realise the need for this integrated approach. In other cases, the need may be recognised as a good management practice when putting together an original tender.

In the author's experience it was found that several man-years of education were necessary before the value of the integrated programme was fully recognised. I found that top-level management were, on the whole, most responsive once the savings in time and costs could be demonstrated. It was necessary to broaden the recognition, by all members of the project team, of the meaning of the word 'quality' from the traditional 'conformance to drawings and specifications' to 'satisfying the needs of the the customer'.

9.4 What is a project quality plan

The project quality plan is an integrated controlled document prepared by a quality engineering professional who brings to the project team a point of view complementary to the other members of the team, such as design and manufacturing. This point of view relates to the assurance, control and verification activities necessary to provide confidence in the resulting process, product or service. This different point of view may identify technical and administrative points in the contract and its supporting documents that may otherwise be overlooked owing to concentration on how the technical requirements are to be achieved. The quality engineer reviews the requirements from the aspect of 'are they measurable or verifiable' rather than 'can we design and make an item to achieve that performance'.

In order to improve visibility on the project quality plan, it is recommended that each plan is given a distinctive number in the documentation series, e.g. a prefix of either Q, QP or similar easily related combination of letters.

Experience has shown that it is best to divide the project quality plan into sections with each section relating to a particular operational function. Each section references or includes the procedures applicable to that function. Subdividing into sections permits each operational activity to have as many copies of that section that are necessary to ensure that all participants are aware of their contributions to the project quality plan. At least the senior member of

- -

ABC ELECTRIC CORP.

Project quality plan

for

(type in project title)

PROVIDED TO

(type in customers name)

CONTRACT NO. . . . . DATE . . . . . .

QUALITY PROGRAMME STANDARD. . . . . DATE . . . . .

SECURITY CLASSIFICATION. . . . .

DESIGN AUTHORITY. . . . .

INSPECTION AUTHORITY. . . . .

QUALITY PLAN PREPARED BY. . . . . DATE . . . . .

QUALITY PLAN APPROVED BY. . . . . DATE . . . . .

REVISION. . . . . PREPARED BY. . . . . DATE . . . . .

REVISION. . . . . APPROVED BY. . . . . DATE . . . . .

page 1 of . . . .

- -

Fig. 9.1 *Typical title page for a project quality plan.*
Certain of the above headings may not be used on some applications, however, the headings are those used by the author. The use of different fonts and character sizes can more clearly distinguish between the headings and the typed entries of the completed Project Quality Plan. However, if word processing is used, the form for the cover can be held in memory and the data entered as required. In this case segregation could be the use of different fonts or bold typing for the headings.

each function would, of course, have the complete plan in order to be fully aware of the integrated programme.

Thus, a particular plan could include applicable selections from the following:

(*a*) Marketing or contract administration
(*b*) Design — hardware and software
(*c*) Material control or procurement
(*d*) Manufacturing or production
(*e*) Customer service including handbooks, instruction manuals etc.
(*f*) Installation or commissioning
(*g*) Field services
(*h*) Quality

- -

page 2 of . . . . . .

ABC ELECTRIC CORP.

Project title . . . .   Project quality plan no. PQP   . . . .

INDEX

SECTION

NUMBER	TITLE	PAGE
	Title page	1
	Index	2
A	Marketing	3
B	Design	
C	Material control	
D	Manufacturing	
E	Customer service	
F	Installation and commissioning	
G	Field service	
H	Quality assurance	

- -

Fig. 9.2 *Typical index page for a project quality plan.*
The above listing of sections would apply to a major project design, installation and commissioning. On other programmes, those sections not applying would be deleted

An index of the applicable sections should be included in the project quality plan. (Fig. 9.2).

Each of these sections can, basically, be prepared in one of two ways:

(*a*) Index form identifying the particular procedures applicable to that section
(*b*) Procedural form where each procedure applicable to that section is included in the plan

These are not mutually exclusive alternatives, since each method can have its application on a particular project.

Internally, every endeavour should be made to minimise the amount of documentation that is peculiar to each project or program. It is advantageous to use standard procedures as far as possible, so that personnel are fully familiar with their contributions and responsibilities. Thus, an index form can be very effective for internal communications. In this form the index would show where a procedure applied as written, and where it may be necessary to modify an existing procedure for the particular project. It has been found very effective to show these amendments as a numbered series of notes referenced in the index.

- -

ABC ELECTRIC CORP.

Project title . . . .　Project quality plan No. PQP　. . . . .

QUALITY PROGRAMME PROCEDURES INDEX

SECTION A MARKETING

NUMBER	PROCEDURE TITLE	Applicable as written	Amended by note no.
A1	Tender (quotation) quality review	n/a	
A2	Contract quality review		A1
A3	New product quality review	n/a	
A4	Work authorization – quality terms	X	
A5	Advertising verification	n/a	

- -

Fig. 9.3 *Typical section index page for a project quality plan*
The above gives a typical quality-programme-procedure index as applicable to the marketing/contract administration activity. The notations in the last two columns convey the following:
(*a*) n/a indicates that the procedures relating to tenders, new products and advertising verification are not applicable
(*b*) A1 indicates that the standard procedure dealing with contract quality reviews is modified by a note A1 where would follow the index
(*c*) X indicates that the standard procedure on 'work authorization – quality terms' applies as written

This results in an index having the headings 'applicable as written' and 'amended by note no.', as shown in Fig. 9.3.

For external communications, it may be advisable to include copies of the actual procedures in addition to using the index as described above. In this case, each operational section of the plan would include copies of those procedures applicable to that activity. This may mean including more than one copy of a procedure where it applies in more than one activity. However, these additional inclusions are justified by preventing the user having to use a single document with possibly more than one modifying note.

These alternatives are based on the following premises:

(*a*) Any new procedures required for a project are included in the general procedures manual so they are available for future programmes.

(b) Operating instructions for particular inspection or test apparatus are not issued as procedures, but as separate instructions or methods data. I fail to see why these instructions should fall into a different category to those instructions issued for manufacturing operations.

The project quality plan should *not* duplicate any parameters or outputs of project activities. Duplication can lead to errors of double specifications due to both specifications not being kept the same. Typical areas include:

(a) Acceptance criteria: These should be included in the drawings or specifications used to define the project output
(b) Schedules: These are the output of the planning section
(c) Methods Data: These detailed instructions may be issued by industrial engineers or methods specialists
(d) Quality cost reports: Targets may be included, but reports are the result of project activities
(e) Project life-cycle flow diagrams: These are an output of project planning

The project quality plan should have a distinctive cover that will clearly segregate it from the standard project documentation. The design and techniques will depend upon the techniques available and the desires of the organisation. The provision of a special title page will usually suffice to provide this segregation and the necessary information. A typical title page is shown in Fig. 1. Additional clarification could be given by the use of different type fonts and sizes.

9.5 When should the quality plan be prepared

The project quality plan should be prepared as soon as planning commences on the project; i.e. the quality engineer should be a member of the initial planning team.

If the project is resulting from a request for tender or quotation (RFQ), a preliminary project quality plan should be prepared to ensure all requirements of the RFQ are addressed. A preliminary plan of this nature will frequently cover the requirements to be covered in the quotation as well as the requirements of the proposed equipment or system. The inclusion of a project quality plan as part of the documentation supporting the quotation can be a strong marketing tool, whether this is required by the RFQ or not.

Once a contract has been received, in response to an RFQ, the preliminary project quality plan must be reviewed against the contract and its supporting documentation, to ensure that any changes are identified before work commences. If any changes are found, these must be reviewed to determine the impact on the resulting product, schedules and costs. If necessary, further negotiations may be required to resolve these. Where necessary, the preliminary project quality plan should be revised and issued as a final document.

If the contract is received directly without a tender or RFQ, a similar review is required to determine:

(a) if the project definition is in clear, unambiguous and measureable terms

(b) what are the quality program requirements

(c) what are the specific assurance, verification or control requirements

(d) what project approvals are required

(e) what assurance documentation does the customer expect to receive

(f) etc.

If a new project is being developed for sale in the general market place, the project quality plan should be initiated at the start of project planning to identify the steps to be taken to satisfy the needs of the intended customers. In this case, many of the more detailed requirements will develop as the project develops. However, the initial plan should ensure the product or service is being defined in measurable terms that can be verified during and at the completion of the project life cycle.

9.6 How is performance monitored

All activities required by the project quality plan should be monitored during the project life cycle to ensure they are being carried out in accordance with the requirements of the plan and with the expected results. This monitoring should be one of the responsibilities of the quality activity, by whatever name it is known.

The most effective method, of carrying out the monitoring, is the quality audit.[8-10] Fundamentally, this technique measures the ability of those involved to make valid quality decisions. It also evaluates the results of the various activities concerned.

There should also be certain control points in the project life cycle that will evaluate the effectiveness of the project. These control points should be project reviews at which the activities and results are evaluated by an independent group of qualified individuals. Independence implies that the reviewers are not directly involved in the project activities. The data to be reviewed should be provided to the reviewers in advance of the formal review meeting, so that the examination can be more thorough than is practical at a meeting. Each review should be charged with reaching consensus on the decision with respect to future actions. Typical major decisions include:

(a) authorization of the commencement of design (feasibility review)

(b) authorization of the commencement of procurement and production (suitability for production review)

(c) authorisation of delivery of the first producer to the customer (suitability for use review)

9.7 Conclusions

This Chapter has shown how the adition of a prefix to the ISD definition of a quality plan can provide an integrated, customised quality plan for a project that collates all of the assurance, control and verification activities into a single cohesive plan of action. The project quality plan complements the production definition or specification by clearly defining the procedures and controls to be used to confirm conformance to the requirements of the specification.

A quality plan of this nature will:

(*a*) Provide visibility on all assurance, control and verification activities required on the project

(*b*) Reduce the risk of errors of omission and duplication in the assurance, control and verification activities

(*c*) Reduce the risk of project delays and excess cost through reduction in errors.

(*d*) Provide a benchmark against which the performance of various activities can be measured

(*e*) Provide assurance to management and the customer that the resulting product or system will achieve its quality objectives; i.e. it will be suitable for its intended purpose

(*f*) Provides an environment for continual improvement of the controls in order to reduce the performance scatter of processes and the resulting product of system

9.8 References

1 MILLS, C. A.: 'Cost-effective quality assurance' *Quality Assurance*, Dec. 1984 **10**, (4)
2 'AQAP-1: Nato requirements for an industrial quality control system' (NATO International Staff, Defence Support Division)
3 CAN3-Z299.1: 'Quality Assurance Program – Category 1' and CAN3-Z299.2: 'Quality Assurance Program – Category 2' (Canadian Standards Association)
4 MILLS, C. A.: 'Management for quality', *IEE Proc.*, Jan. 1983 **130**, Pt. A, p. 000
5 ISO/DIS 8402: 'Quality assurance vocabulary' (ISO, Geneva, Switzerland)
6 ANSI/ASQC A3: 'Quality Systems Terminology' (American Society for Quality Control)
7 BSI 5750: 'Quality systems. Part 1 Specifications for design, manufacture and installation' (British Standards Institition)
8 CAN3-Q395: 'Quality audit' (Canadian Standards Association)
9 JURAN, J. M.: 'Quality control handbook' (McGraw-Hill)
10 SINHA, AND WILLBORN: 'The management of quality assurance' (John Wiley)

Chapter 10

Computer applications

D. Langan

Director, Taylor Woodrow Services

10.1 Introduction

Since the first significant discovery of oil in the Forties Field in late 1970 a considerable number of large multi-discipline projects have been organised to implement investments for the recovery of oil from the North Sea. Each project involves the installation of one or more large support structures on the sea bed in depths of water up to 180 m as base for a range of services installed in the form of prefabricated modules. Such modules are densely packed with many forms of equipment and can be the size of a small hotel. Projects typically have a scheduled timescale of 3 or 4 years with average expenditures of up to £1M for each working day.

With such exceptional physical and financial scales one key factor is the timely collection, organisation and use of vast amounts of project information for monitoring and control purposes. This Chapter is concerned with our experiences in this area of project activity.

Characteristics of North Sea oil projects that are pertinent to the design and implementation of information systems can be summarized as:

(i) The need to mobilise, train and structure large multi-discipline teams of people with widely differing experiences over a short period of time

(ii) The very strong necessity to adhere to schedules because of the seasonal nature of the weather and the resultant sea conditions (weather windows) so as to minimise the time from the initial investment decision to receipt of first oil revenues

(iii) The use of geographically widespread locations for the supply of materials/equipment and the provision of fabrication services for modules and elements of the main platform structure or jacket. There is thus the need for the complexities of currency-variance measurements and multi-currency reporting.

(iv) The necessity to overlap design with procurement and construction activities in order to reduce the overall duration of the project.

(v) The need to monitor and track all components for both construction scheduling purposes and to provide traceability for safety certification, insurance and maintenance purposes.

(vi) The occurrence of design changes and associated material/equipment requirements necessitates the facility for rapidly assessing the impact of these changes through all stages of procurement and construction.

Any one of these characteristics may not be unique to North Sea Oil related projects, but in combination they create a singularly complex environment.

Allied with the technological advances and the acquisition of organisational expertise has been the development of information systems for the management and control of a wide range of commercial functions as well as the logistics of assembling large fabricated components and all the equipment within the many types of modules.

Whilst in some cases project control systems have been based on time-analysis networks, our own experiences have been much more concerned with systems for commercial and materials control. This is not to say that network analysis has not been an important function on projects with which we have been connected, but that the systems have been developed to respond to a level of detail which will handle individual material and their corresponding financial transactions.

A critical factor within the project environment has therefore been to create a level of management awareness of the many facets of information available and a resultant control that can be exercised in order to achieve the project objectives. Our information systems, therefore, have been designed to make an important contribution to the efficient running and orderly progression of such projects.

10.2 Scale of North-Sea project information

Major projects such as those for the North Sea produce extremely large volumes of data, and without computer-based information systems the times taken to monitor and provide management information would be well beyond the response times needed for effective control of the project. These systems are also the only practical way of storing and controlling such information for effective use within the project.

Because of the size and diversity of the project the expenditure is very high over a period of several years, and therefore requires a wide range of accountancy controls. There are also tight time scales related to this high capital investment, so that tiered networks for reporting at different levels of responsibility are essential.

From our own experience Table 10.1 gives typical quantities for the number of records of information which have been processed and stored for a North Sea Oil project.

Table 10.1

Information	No. of records
Material requirements	150 000
Purchase orders	5 000
PO line items	48 000
Shipment details	100 000
Material issues	150 000
Steel-plate	100 000
Document control	60 000
Bought ledger	25 000
Financial-accounts history	400 000
Cost-appraisal entries	150 000
Budgeting details	30 000

10.3 History of development

For the Thistle A project a Taylor-Woodrow/Santa-Fe joint venture team was employed in 1974 as the project manager. We already had adequate systems for management of large projects, but these were generally not applicable to the particular demands of the offshore business described above. During the contract, therefore, various attempts were made to design and introduce further control systems. Embrionic systems for monitoring the procurement and logistics of materials and steel were tried and were partially successful in the area of instrumentation during 1975.

Whilst the direct benefits were limited, a great deal of awareness was created on how systems should be introduced, as well as the pre-conditions necessary for successful implementation. Because of this increasing awareness of the importance of information systems to our business of controlling and supporting projects, on behalf of major clients, a crash programme of development was undertaken in 1976–77. This culminated in Taywood Santa Fe (TSL) implementing systems for the North Cormorant project for Shell UK Ltd. in 1978, which then ran continuously until project completion in 1981.

Starting in August 1979 the systems outlined in the paper were used operationally for about four years on the Marathon Oil Brae A project, and were considered to be very successful by the joint project team.

The systems originally evolved during this 8–10 year experience of large projects were written entirely by Taylor Woodrow and used in their developed and modified form until 1983. At this time it was felt they should be further enhanced and made much more interactive. In addition, we considered it essential to use a database system for the technical management of the data within the computer; the DBMS Adabas was selected for this purpose.

A further generation of the system was then written by a joint team of Marathon Oil and Taylor Woodrow with the costs of the development shared. At this time our joint experiences were incorporated into the business requirements. This final version took two years to complete and contains nearly 1000 programs; the final operational system is therefore very comprehensive and sophisticated.

10.4 Systems development and operating environment

At the time of the final joint development, both Taylor Woodrow and Marathon Oil had IBM mainframes with database software to support the size of project for which the systems were required. This hardware was therefore selected, as both that and the software environment were undoubtedly going to develop and, more importantly, were going to be supported by commercially stable suppliers for many years to come.

Early in the design stage, the possible use of minicomputers was considered. However, our experience of the data volumes and processing requirements generated by a major project suggested that this was not a viable alternative to mainframe-based systems.

The mainframe offered the best range of facilities with its sophisticated operating system and powerful database management system. Perhaps, more importantly, we believed that very extensive functionality should be designed into the system, which at the time of the design demanded the use of a medium to large computer.

However, it was expected that the existing trend of significantly faster and more compact processors producing lower heat output would continue. Thus we could design for hardware of the future such that, in a relatively short timescale, processors then available would be physcially smaller and require less stringent operating conditions, i.e. a self-contained processor operating essentially in an office environment.

The systems are currently running on a large central IBM computer with subsidiary processors and terminals located in widely spread locations in Western Europe, effectively supporting a major North Sea development.

10.5 Outline of systems

Very briefly, the integrated systems support the following disciplines for monitoring and controlling a large project

Budget and estimating	Planning
Cost budgeting comparisons (Cost appraisal)	Materials control
Cash-flow forecasting	Steel Control
Payments and financial accounts	

Each system can operate independently within its own functional area, but considerable further benefits accrue when information cross flows are permitted (i.e. full integration). In this case, the systems are represented as in Fig. 10.1. Integration allows information to be inputted in one area for multiple use in different combinations in other areas of the project. The use of database techniques was a further step towards making full use of this principle.

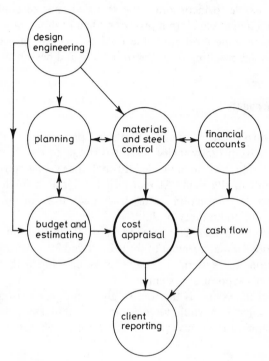

Fig. 10.1 *General interface between systems*

10.6 Budget and estimating

The function of the system is to store and control a large volume of estimating data from which budgets can be produced for oilfield and other major construction projects; these can be revised by inputting historical data as well as new information such as changes in scope or improved estimating data based on the latest judgment of productivity rates or actual costs from the cost appraisal system.

It is possible to operate a number of budgets within a structured cost code across a number of projects within the same system. These budgets are built up by accessing the same central data including inflation indices, exchange rates and resource rates, which can all be regularly updated. Budget values are spread through time in accordance with a variety of phasing curves stored in the system. Inflation indices are then applied to the phased values, different indices being used for different types of resources or alternatively groups of resources. The system is also able to report inflation amounts due to changes in timing, prices and rates since the start of the project. The project base data value represents the cost of the project without inflation.

Finally it is possible to incorporate the required level of contingency, e.g. lump sum or percentages etc., into a project budget at different levels of detail. By use of the system the most probable final cost of the project is forecast, together with a range resulting from the optimistic and pessimistic predictions.

10.7 Financial accounts

The system provides a means of controlling and processing all project-related invoices and cost from the point of receipt through to payment. On receipt, invoices are entered to the system, passed through the various stages of checking and approval, and finally paid. At many of these stages in the system it is possible to report on transaction status, trace back to source documents and provide control and auditability of data.

Some 20 reports are available to provide financial information relating to the items handled both for management of the project and for government statutory rating. In addition, the system is capable of interfacing with the materials-control and the cost-appraisal system.

The purpose of the system is threfore to give a far greater degree of control and reporting facility than would be possible in a manual system, and, bearing in mind the volume of data involved, greatly improve efficiency.

10.8 Cost appraisal

The function of the cost-appraisal system is to accumulate all the financial information of a project, the reporting of this data in various formats, and the projection of final cost.

The system provides a method of monitoring the financial position of contracts and purchase orders. Purchase-order information is derived from the materials system, whilst contracts are controlled entirely within the cost-appraisal system. Figures are held at a detailed contract and purchase-order level for commitments and invoices (the latter being part of financial accounts).

Costs are reported by project-cost code, to facilitate comparison with the original budget (control estimate), or subsections of the budget authorised for expenditure (AFE), which provides the comparison between budgeted and actual expenditure.

The effects of price inflation since project start date, and fluctuations caused by exchange-rate variation for those purchases/contracts paid in foreign currency, are reported separately to allow the 'true' variation between budget and actual to be highlighted. Financial data assembled by the system are utilised to give the best estimate of immediate cash requirements and a forecast of cash flow to the end of the project.

10.9 Materials and steel control

The materials-control system provides information commencing with the design stage and through all aspects of procurement and distribution; the as-built use of materials is then recorded against the original material take off (MTO). The steel-control system provides detailed information on the availability and use of individual steel plates and plate products.

(i) *Material Requirements*: The system is designed to accommodate requirement information at bulk and detail levels against design locations. The system automatically accepts detailed requirements against a design drawing in preference to bulk or preliminary material take-off (MTO) details.[1] This information is held by material identification number (MIN) on the requirements file with a description of the MIN which forms the basis of a Materials Catalogue.

(ii) *Order/Expediting*: Once a requirement is defined the system can produce details which prompt the purchasing group. Any orders placed are then fed back to the system and recorded on the orders file. This enables the purchasing group to control the level of purchases and to ensure that all requirements will be met. Expediting data, details of shipments to the various fabrication sites and the resultant goods receipt notes are also held on the orders file.

(iii) *Goods issue*: As material leaves the stores, it is recorded on a goods issue note and entered on the system. This provides the second half of the stock-control facility and, provided the issues are recorded against specific areas of work, the system will check that issues are in accordance with the original MTO.

Control of steel up to distribution is also handled through the materials system, but, after that point, detailed records of all steel items, offcuts and pieces are maintained on the steel control system.

10.10 Planning relationships

The flows of information between planning, materials control and financial reporting are very important; some aspects of these relationships are briefly described below.

The systems were designed such that a number of planning systems could be interfaced with the total system. This approach was taken for a variety of reasons:

(a) A fixed link to any specific planning system was not desirable in the contracting environment, as prospective clients would invariably already have a preferred system. Thus only the ability to provide an interface was considered.

(b) planning dates tend to be used as guidelines only in the procurement activity, as the manufacturing/shipment/customs lead times all need to be considered and combined before an order can be placed.

(c) Changes in plan should not automatically change procurement requirements, but primarily the dates by which materials are required. Changes in plans and schedules will often not alter the physical material requirements, but rather the destination to which the manufactured articles are shipped and the time they are required. The system concept employed is that materials, when available, are shipped to the location where they are currently required rather than where they were ordered for.

The planning information required by the materials and financial functions are obtained via the MTO, and this is the vehicle for the transmission of the relevant data — primarily material and services requirement dates. Materials procurement attempt to met all required planned dates, and what cannot be met is reported manually to planning via exception reports.

From the moment an order is placed for material or a contract for services of any kind, the financial aspects of that transaction are automatically passed through to the financial side of the total system and the impact on both liability and cashflow are assessed and reported.

Although only a manual interface is currently defined within our procedures, the details provided by the MTO system, together with the associated planning details, are passed through to the budget and estimating system. Initially, this information is at the preliminary level to support the creation of the budgets; subsequently as the MTOs become more refined, it is used to update and improve the detail within the budgets.

Thus the cost impact of both the uncommitted portion of materials and services from the budget system, and the committed portion from the materials and financial account systems, can be assessed through the cost-appraisal system at any stage of a project. This supplies both funding and cashflow-requirements details and provides project management with the information necessary to assess the financial implications of the planning activities. Through this facility, and the ability within the budget system to operate alternate 'what if?' budgets, financial constraints on the project's funding can be passed back into the planning activity.

Finally the financial details of materials received are used to automatically calculate the valuations of stock in hand in the inventory control system.

10.11 System design problems

Whereas corporate or manufacturing systems are designed for a fairly continuous production environment, systems for project management have to be adaptable to different contractual situations, a range of clients and to interface with their corporate systems.

Clients often have completely differing views as to how the project is to be run, and therefore the systems must be able to accommodate a range of business principles. For example, on the majority of North Sea Oil projects materials and equipment are often procured by the client's representative for free issue to the contractors. The purpose of this arrangement is to shorten the overall time scale of the project, whilst placing greater emphasis on central control. This has a large impact on the control of the procurement and materials-movement functions. Instead of managing at the work package or sub-contract level, all items of material and equipment have to be recognised and monitored from the time when the requirement is first established at the design stage through to the 'as-built' location in the construction.

Large construction or fabrication projects are influenced and conditioned by the client's own organisation and his available resources, together with the market conditions which affect contractual arrangements. The location of fabrication sites and the size of the physical structure to be constructed, together with the constricted time scale and the lack of available float, all dramatically impact the degree of difficulty with respect to systems implementation.

Two types of contract occur in practice: project management, where the contractor manages all aspects of the project for the client, and project support services, where the contractor provides a range of skills and services which are integrated with the client's own team.

In both cases a large group of people are brought together from many disciplines and backgrounds. The project manager's systems are more generally used in the first case, whereas, with support services, considerably more diplomacy and discussion are required to establish effective operational use of systems by both the managers and client's teams.

Whereas the project manager is mainly concerned with accounting at project level, the client has procedures and systems which account for his overall business. Our experiences have indicated the need for an interface to transfer details to and from the client's corporate systems and the project-management system.

Project-management systems must therefore have great flexibility, both in terms of being able to respond to different business principles and project organisations, as well as the ability to interface with the corporate systems of a range of clients.

This requirement for flexibility to meet varying management styles increased the overall cost of the systems development. However, the resulting system benefited from this additional effort, which undoubtedly improved the product and has provided a more adaptable tool for the contracting environment.

Development costs were reduced to some extent by the decision to provide users with (limited) facilities in a fourth generation language, thereby providing them with the ability to develop reports for their business area, over and above the main core of reports included in the development.

Wherever possible, strict philisophies have been avoided and the implementation of the systems in differing environments is catered for by providing:

(a) the ability to write flexible corporate interfaces with the client's existing systems

(b) the ability to have coding structure for materials identification, and on the financial side costs/budgets/currencies etc. which were initially undefined. The codes and associated dictionary of terms were then set up at the beginning of the project to suit particular requirements.

In addition, there is a very extensive list of menu items for calling up transactions and reports; these items are customised to each functional area of the project, as well as the level of responsibility of the prospective user.

Our success in the development and use of project-management systems has depended on building up a range of experiences which demonstrates an established mode of operation to both the clients in a general sense and specifically to individuals joining the project team.

10.12 Early experiences

The following were typical situations and experiences during the early development and implementation of our systems in a project environment:

(a) Users did not clearly know what they wanted or how a computer-based system could really help them, and quite often solutions being requested would not resolve their underlying problems.

In some cases the user was most concerned in trying to solve problems relating to his most recent experiences which were conditioned largely by the recent past. Difficulties in communications did not help to foster understanding between the systems designer and the future user, and problems were often generated by the people concerned making assumptions on the other's level of knowledge.

(b) Initially users had difficulty in appreciating basic computing concepts, e.g. the differences between commercial and scientific/engineering applications, which is essentially the difference between personal computing and multifunction systems. In the latter case there is multiple input from many sources and multiple output combining this input in different ways. Each user, therefore, depends on a number of others as well as himself for his information from the system. Because the input is from multiple sources, extensive validation is required, as the people performing this routine task may have no particular interest in the subsequent use of the information.

(*c*) As each discipline defined its needs for the project, this identified the various sources of information, which in some cases was available from more than one location. The system designer then established the most appropriate source for any particular piece of data. This resulted in an interdependence between disciplines, with users having to be convinced of each others performance and timeliness. This rationalisation of input also led to imbalance between the volumes of input required from one particular function and the resultant benefits to that area from the use of the system.

In some cases there was little or no benefit at all. For example, the design or engineering function were defined as the source for all materials take-off details, but they received little benefit from the system.

Our experiences suggest that, to successfully motivate a range of users a concentrated effort in preparation of detailed training sessions covering the whole scope of the system is essential.

The training of *all* users in detail for their particular tasks and functions, as well as in general to illustrate how their role fits into the overall process, proved to be highly successful in motivating accurate and timely input of information. This is backed up by a help-line facility where the now-known trainers are available to provide assistance when needed.

At design time, extensive care was taken to ensure that the data requested from a particular user, although not necessarily used by him, would, however, be easily accessible for input and generally available to him throughout the project.

The flexibility of both the system and the access to its facilities via tailored menus provide the ability to direct the input of data to the most appropriate source, depending on the management style used by any given project or client.

Finally, the design of the systems heavily involved the users; they were closely involved at each stage of design and development, and from these early stages appreciated that the systems being developed were theirs and that their requirements were being satisfied. This engendered a feeling of ownership and involvement, which undoubtedly generates a desire to make the system work.

(*d*) The introduction of systems into the complex organisation of a major project required a great deal of commitment on the part of senior management. This was important, and needed great resolve, particularly during the implementation of the system on the first or even second project. Allied to this management pressure was the need to demonstrate a few key areas where a manual system of control was clearly beyond practical limits. Individuals had variable resistance to systems, and in some cases they believed that system designers were telling them how to do their particular task when they had been hired for their experience and expertise in that area. In addition, people newly assigned to a project had preconceptions of the procedures to be used and did not wish to spend time assimilating the way systems could be most effectively used for the new project.

(*e*) It became clear that project control systems had to be sufficiently adaptable to operate within many different business environments agreed with the client or introduced by the management contractor. Individual project procedures were then defined to implement this business philosophy on a day-to-day basis.

(*f*) For project-services contracts a front-end procedures study was a very effective means of establishing how the information and control systems were to be adopted for the project and interface with the client's corporate procedures. The results of this type of study must be disseminated and fully publicised within the project team so that systems implementation gets off to a flying start in parallel with the conceptual design stage.

(*g*) One of the requirements of North Sea projects mentioned in Section 10.1 is that of overlapping design with procurement and construction activities to reduce the overall time before revenues are generated.

The North Sea environment in the late 1970s and early 1980s was such that the construction technology was developing almost as fast as the computing technology. It was inevitable that major revisions of rig design would occur until quite late in the construction period, both onshore and offshore.

The systems have therefore been designed to operate effectively within this changing construction environment. Such a requirement has not caused any limitations in the design of the system, rather the reverse, as additional functions and logic have had to be introduced so as to support conditions actually occurring in practice.

A particular example to illustrate this flexibility was the severing of the correlation between a material requirement and a purchase order. As previously mentioned, the approach followed was that materials will be shipped to where they are needed at the time they are available, rather than to where they were initially ordered for.

The system provides a comparison of requirements in relation to materials procured at the project level; as the requirements based on the latest design information become available, the final detailed materials take-off is equal to the materials procured.

(*h*) The adoption of database techniques were considered from two aspects:

    (i) benefits from a user's viewpoint
    (ii) aids to the designer.

The user obtains benefit in that the associated 4th-generation language provides a fast and relatively simple method of producing reports or inquiries from the data retained within the database, thus providing the ability to create outputs that were not envisaged during the systems-definition stage. The techniques employed when developing the database also ensure that data are collected as near as possible to its source, and is only input once, with the result that all users, even though they may have different access and use of the data, are assured that the information they receive is the same.

For instance, a materials controller in the head office knows that the details for a goods-receipt note for materials delivered on a remote site are the actual details entered on that site, and that the information he sees is up to the minute. The controller has not needed to wait for a piece of paper to provide that information.

Since the systems were designed for a wide range of type and size of project, the varying requirements of clients may at some stage necessitate minor changes or enhancements to the systems and the data held. The database approach provides a simple and effective method to achieve these changes to the storage of data without the need to change existing programs which access the files. Only those programs which need to recognise the new information need to be changed.

Additionally, the use of database provides an easier route to enable the extension of the suite for completely new areas of business.

10.13 Conclusions

In conclusion, the benefits available from the use of such systems described in this paper are considerable; in this respect it is important to recognise:

A. First-time use of any systems will only produce limited benefits, and indeed will highlight may other problems. By successive use of the systems, increased returns will accrue as their acceptance increases and the overall management/ organisational structure is established to ensure that the systems are utilised so that the maximum benefits are achieved.

B. The almost revolutionary pace and extent of improvements in hardware and software now results in the facility to provide the users with a basic system, and to allow them to develop some of their own additional programs for *ad hoc* reporting. This leads to a certain level of computing skills being added to the other professional capabilities of the user, provided that there is the interest and lack of prejudice. There is potential for further step-change improvements in computer hardware; software design must therefore recognise and take account of this phenomenon. To provide even greater flexibility for selective retrieval of information, database techniques have been incorporated in an improved version of the systems described in this Chapter.

C. For successful implementation of project-control systems many forms and levels of training/education to present the advantages and effective use of systems must take place. This has to be a continuing process and have the commitment of senior management, as the activity directly competes for the professional's active time on the project.

Chapter 11

Essentials in project management

Harold Masding

11.1 Introduction

First, project management is not an end in itself but a means whereby a commercial production or process plant is installed and set to work as quickly as possible thereby earning the return required from the investment. Success can only be judged by the achievement of timely operation of the plant and the containment of the cost within the budget. The principles of project management apply to many forms of plant investment, and the example drawn in this Chapter is that of power-station design and construction which is highly capital intensive and covers a wide range of interdependent high-technology plant and services.

The cost of delay to the project overall is probably the most significant area of overexpenditure likely to arise and the real objective of management is not to provide a vehicle for solving all the industrial, environmental and industrial relations problems that obtain, but to ensure that the committed investment is completely and commercially secure. Fig. 11.1 shows the incidence and total expenditure typical of recent power-station construction, including the inflation and interest accruing during construction. The effect of an underestimation of 20% and a delay of two years virtually doubles the sum originally considered at the time of placing the order,[1,2] and the most important responsibility placed on project management is the duty to maintain the programme and budget. It is essential that this responsibility is not confused or diluted by involvement with aspects of design or policy outside the project manager's competence, especially during the construction work itself. The view taken in this Chapter is that the only real objective is the commercial success of the project and the need for single mindedness and strength of purpose in the management of the project.

11.2 History

Between 1961 and 1965 forty seven 500 MW units, about 23000 MWso, were ordered and substantially completed within the decade or so.[3] This was

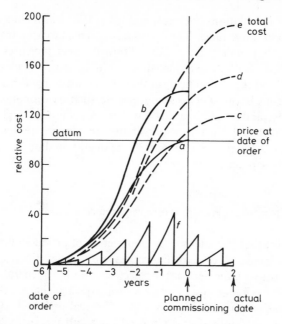

Fig. 11.1 *Incidence of expenditure*
  a Incidence at order-date prices to programme
  b As *a* plus inflation and IDC
  c Incidence at order-date prices plus 20% extras and 2 years late
  d As *c* plus inflation
  e As *d* plus IDC
  f Annual payment if self financing (excluding IDC)

equivalent to an average installation rate of about 2500 MW per annum after a five year lead time. The shortest time to complete the first unit in these stations was 51 months, the average was 70 months and the longest over 98 months. The final costs, excluding interest during construction (IDC) adjusted to the same price level at 1971, varied from £55/kW to £75/kw and the increases over and above the original estimates from all causes except IDC were between + 10% and +25%.

The history of the construction of these stations was examined by the Wilson Committee in 1969.[3] It is considered in hindsight that the wide variations in station-design practice and in project management were also contributory the poor results and insufficiently highlighted at the time.

More recently, an expert group set up by the Nuclear Energy Agency (NEA) of the Organisation for Economic Co-operation and Development (OECD) published in 1983[4] a comparison of the power stations built or proposed by utilities in Europe, North America and Japan. Great care was taken to create a common basis for comparison which showed that, for both nuclear and conventional power stations, the nominal cost in the UK was about twice the of some similar installations overseas, and when IDC was included, this factor increased to about three times. The excess costs seemed to arise from some high prices initially, but mainly from expenditure to meet increases during construction and IDC. Both these latter aspects point to project management as the high cost area, and the real indication that should arise from this review is the identity of the areas on which to concentrate such as economic design and effective buying, and to eliminate those aspects leading to high cost, delay, diversion of effort and blurred responsibility. It is worth reiterating that the criterion for success is the achievement of the operation date at a cost comparable with the best practices overseas.

11.3 Existing plant margins and future requirements

One other aspect that has affected the completion of power stations in the UK has been the absence of the consumer pressure on the utility, caused by the ample margin between the demand for electricity and the generation capability and the remote likelihood of power cuts. It is expected that, in the late 1990s, the ageing of the existing 500 MW and 660 MW coal-fired and AGR plant will reduce the plant margin and create a need for a substantial replacement programme lasting for several decades. There are other considerations such as the practicality and economics of major refurbishments of existing high--performance plant which could delay the replacement of some of the time expired stations; also the possible installation of new low-fuel-cost plant in advance of system need for economic reasons. Even so, the industry's future work load will be considerable and will emphasise the need for a continuous ongoing manufacturing and site construction programme based on the replication of single unit designs.

If, for any reason, the nuclear construction programme was deferred, it would be essential to develop and install more efficient, low-running-cost, base-load thermal plant utilising modern technology with supercritical steam conditions, but this aspect is considered to be outside the context of this Chapter other than to assert that the installations would be based on a replicated single-unit design and the project work load would be of a similar order.

With regard to the future generation needs, the first essential is that the fuel policy, i.e. the effective allocation of plant to burn coal or nuclear fuel, is decided within the next decade, taking into account the availability and cost of the fuels, the acceptability of the process to the environment in all aspects, the national resources, the economics within the industry and the tariffs forecast for the consumers — and avoiding any significant duplication of capacity. From this policy will come the timetable and the numbers of the new generating units needed to meet the likely replanting programme and growth if any, in good time permit realistic updating of the quantities involved before commitment to al work.

The second essential is to have available fully developed and competitive designs for the boiler (nuclear and coal), turbogenerator and auxiliary plant ready for commercial installation in the UK in economic and practical stations devoid of waste and unproductive elaboration — preferably 'skinned to the bone' and designed for replication as single units.[6] The construction programme postulated in Section 11.5 suggests that the utility could then order both coal and nuclear units 'in bulk' and call off its annual requirements with real economy to both purchaser and contractors, provided that the industry can create an economic design suitable for replication and achieve an acceptable construction timescale. This latter task is the duty of project management, and with this availability of economic designs suitable for replication, the terms of reference for the project manager can be limited to the investment planning and the achievement of completion. It is this concept that provides the basis for the opinions expressed in this Chapter.

11.4 Decision to proceed

The key date for project management is the date of the decision to proceed with a particular installation. The detail of the plant design, the allocation to a site, the station design and layout, comparisons of alternatives, investment needs etc. can and have already taken place without any major financial commitment or actual expenditure. On the definite instruction to proceed, there follows a firm commitment to purchase the unit boiler and turbogenerator plant, and expenditure on a large scale begins to flow — the project is indeed a major investment and competent management is essential to get the return on the investment. Project management does not have to concern itself with the reasons for the station or attempt to amend or justify any of the preceeding decisions — it is entirely concerned with building to time and cost. The belief that project managers are infallible and can interfere at random should be discounted completely.

11.5 Summary of work completed before authorisation to proceed

The concept of project management in this Chapter is that of the supervision necessary to construct one of a number of replicated units or stations over an on-going period up to 20 years or more. The following areas of design and cost policy have already been decided and are outside the project manager's competence to alter in principle or in detail. All pre-start experience has been fed back into the pre-contract stage, and the only significant alterations during the construction will arise from rectification.

(i) Date of full commercial operation, the load factors and system economic data applicable over the lifetime of the plant; the fuel, type and qualities expected and designed for, the quantities required and sources.

(ii) The size of the unit and the technical capability, the key areas of standardisation and replication, the definition of any approved advance in design and technology and the associated proving programme (back scheduled to meet the operation date); the main building layout, design of foundations and superstructure.

(iii) The ultimate capacity of the site, the space to be retained for future developments and the practical mode of building on site, the detailed site layout for the particular project.

(iv) The water sources, quality, quantity, availability, storage, treatment process and system needs for the boiler-feed make-up water system.

(v) Similarly for the condenser cooling-water system, including the hydraulic gradients, pumping power and head optimisations, interconnections; cooling towers, pump type and details, filling, emptying and operational aspects, location of make-up and purge system, location of intake and outfall works and detail designs.

(vi) Off-site ash disposal areas, locations and capacities, basic concepts, planning conditions and system designs.

(viii) Access for nuclear fuel delivery, storage, handling, decontamination, moniloading facilities, bunker feeds and overall system designs.

(viii) Access for nuclear fuel delivery, storage handling decontamination monitoring, waste-disposal system design and management-control system designs.

(ix) Transmission connections − basic system design, feeder and generator connections, interconnections and protection, location of incoming lines and terminal towers, orientation and location of the switching station on site, line and cables routes, line clearance and ground sterilisation and installation programme. The auxiliary electrical system design, loadings and protection.

(x) Off-site access roads for construction, on-site roads, site storage areas, men and materials circulation for construction and the permanent works.

(xi) Plant interchangeability, stores standards, materials approvals, tests at works, 'competent' approvals and inspection of pressure parts, welding procedures, tests and records at works and on site.

(xii) General and detailed site particulars, topography, levels, cut and fill balance, sub-soil bearing capability, amenities values and commitments, tentative civil foundation designs, ZVIs, aircraft obstruction, coastal protectional measures, flood levels and probabilities, radiation monitoring and control, mining support and subsidence, settlements and protection.

(xiii) Consent conditions − details of all applications, approvals, stipulations and agreements reached with all statutory and other competent bodies, mandatory actions to be implemented.

(xiv) Storm damage, flood protection during construction and for the permanent works, foul drainage, treatment and disposal and consents.

(xv) Site investigations, definitive foundation design, location of main and ancillary buildings, temporary works, security areas and provisions during construction, commissioning and operation, mode of development, access, clean change and amenities for the operating staff etc.

(xvi) All technical and contract drawings for the inquiries and accepted tenders, submissions for approval and approvals from the specialist engineering departments.

(xviii) The commercial strategy approved by the purchaser.

(xviii) The manning levels and extent of automation.

(xix) The estimated final cost of the project including provisions for contingency, price escalation and interest during construction. The estimated final cost projected throughout the project is probably the best indicator of the likely success and *must* include estimates of the future escalation and IDC.

In a programme of work as big as that envisaged in the replanting programme, the replication of the selected boilers and turbogenerators and the appropriate building structure and superstructure is essential. A continuous investment programme lasting two decades allows the purchaser to call off the delivery requirements over the period and with the least possible disturbance to the manufacturing programme. The start date for each project in programme would be the date when authority was given to call forward the plant already provisionally contracted for. The plant loadings on the structures and foundations and space details will all have been settled in the initial definitive studies, and the outstanding work for each project would be to amend the foundation detail design to suit the actual ground conditions, settlements etc., for the particular site; these changes should be minimal without any need for design contracts.

11.6 Estimated cost, contingency and commercial strategy

The estimated cost of the repeat units and stations would be established by the definitive orders for plant and services for the first stations updated to suit the date when called forward, to which would be added the cost of the civil works for the particular site. This information should provide firm price information for at least 80% of the total cost of the works, and there should be little need for a contingency in excess of 3–5%, certainly not that approaching 20% as indicated in the MMC and OECD reports.[2,4]

The commercial strategy would be decided by the purchaser before giving the authority to proceeed. By 'strategy' it is meant the selection of the method best for the purchaser to buy the plant and services. This may change to suit the 'climate' at the time of purchase. It is expected that, whatever method is selected, the boilers, turbogenerators and main auxiliary plant will be purchased from competent manufacturers on a design and manufacture basis with erection on site and setting to work limited to the manufacturer or to an equally competent experienced contractor. The contract strategy will determine whether to purchase the rest of the works through a multiplicity of direct contracts (as in the 1960s programme) placed after competitive tender, with the purchaser's own engineering departments providing the system designs, planning, co-ordination

Table 11.1 *Options in contract strategy*

Work area	Purchaser's contract with		Type of contract
	Main contractor	Sub-contractor	
1 *Unit/station design*			
Steam cycle and main plant,	Directly employed staff	Specialist constrs. and	% contract price
CW system	Consultants	plant manufacts.	Fixed fee
Aux. & services	Architect Engr.		Cost reimbursible
Planning, co-ord, and	Managing contractor		Cost reimbursible with targets
progress	Consortium		Negotiable
Engineer's approval	Agency or joint venture		Undisclosed (in overall price)
Contract management			Turnkey, joint venture
2 *Plant manufacture*			
Boiler and aux.	Specialist D & B contractor		Lump sum, competitive
Turbogen. and aux.	Consortium	Specialist	Single tender, negotiated
Ancillary plant	Architect engineer	D & B contractor	Undisclosed within price
			Cost plus with/out target
3 *Plant erection*			
Boilers and turbogens.	Direct labour		Lump sum, competitive
Ancillary plant	Spec. D.&B. contr.		Cost plus with/out targets
	Spec. erection contr	Direct labour	Competitive rates and
	Architect eng., managing contr.	or spec. subcontr.	remeasure
	or consortium		Turnkey – undisclosed in
			overall price

Table 11.1 *Continued*

4 Design of civil works

Specialist consultant	As for 1
Contractor, managing contr.	
Architect engr.	
Consortium	
Agency or joint venture	

5 Construction civil

Direct labour		Lump sum, competitive
Specialist contractor		Compet. rates and remeasure
Architect engineer	} Spec. contractors	Cost plus with/out targets
Consortium	} or direct labour	Turnkey, undisclosed in detail

and the direct management of all contracts, or to place one or a few comprehensive contracts for large areas of work, relying on the main contractors to
co-ordinate the smaller works in detail and manage their sub-contractors:
whether or not to use external consulting engineers to carry out the engineering
functions in part or *in toto*, to use managing contractors, whether to use
competitive bidding throughout, in part or to use negotiated contracts in
particular areas with little or no competition, lump-sum contracts with or
without price-increase recovery, cost-plus contracts with or without 'targets'.
The compromise lies between contracts for small parcels of work, fully defined
and priced but needing a lot of interfaces and detailed co-ordination (and scope
for going wrong), as against larger all-embracing contracts with less definition,
co-ordination by the contractor, fewer client interfaces, the possibility of con-
firming the details as the work progresses and including last-minute changes
without wholesale re-pricing or delays. Generally the claim has been that large
contracts tend to be more effective, as more goodwill and sub-contractor control
is involved, but this does not apply in a delay or loss-making situation when the
real pressure reverts to the client to control or make good the losses.

The objective is to secure the best technical and commercial terms whereby
the purchaser may achieve the operation dates at the lowest final cost. Differences in station or plant performance should be insignificant whatever the
form of contract, because the heat cycle and plant is near enough the same in
all cases. The difference is in the station design and the ability to complete to
time and cost: the choice in contract strategy really concerns the civil and
ancillary plant works, and the likely financial gain from open competition on the
one hand or better security of timely completion on the other arising from better
co-ordination and incentives, with the allocation of the small works to associates without competition. The alternatives available to the purchaser are set out
in Table 11.1, and again it is considered worthwhile to restate the objectives of
the contract strategy:

(*a*) to buy plant that achieves the required performance and availability
(*b*) to achieve the planned commercial operation date
(*c*) to achieve the lowest total cost
(*d*) to use all the technical and commercial skills available in the national
      interest.

11.7 Planning and programme requirements

The first essential is to set out a realistic but 'tight' construction programme
based on the practical sequence of work without contingency and clearly
defining all interdependent design and construction stages, the precise dates
when the design information has to be finalised, certified and exchanged with
other contracts, the dates for the start of manufacture and delivery of all major

components, the access dates and intermediate completion dates for all sections of the work down to each column foundation. Both critical-path networks and bar-line charts have advantages and should be used as and when convenient: the best planning systems use both effectively. It is more important to include the judgments of responsible management than a lot of detail derived from assumptions at a junior level. The overall situation and the more complicated sections of work can be sorted out by computer techniques, but this is best left to the in-house project specialists. It is not necessary to make every 'small' contractor comply with an elaborate and expensive computer reporting system, provided the basic progress information is available and updated systematically. The stations completed in the shortest time in the 1960s programme did not depend on computer monitoring of networks, and benefited from simple and direct communications. Too much money is spent on maintaining unproductive systems and staff.

The information thus built into this overall 'master' plan becomes the key contract information for access, start and completion dates to be agreed and built into the individual contracts, whether separate direct contracts, subcontracts or sections of identified work in large comprehensive contracts. Without doubt, the contracts best for both purchaser and contractor are those in which the work content can be properly defined and carried out without interference from adjacent works or by continuous amendment in detail. Changes in content, access and time affect the contractor's costs, and the reimbursment claimed and are not helped by unquantified generalisations or agreements to agree in the initial planning stages, or at any stage for that matter.

11.8 Advanced technology

The last point regarding contract strategy (and hence project management) is the need to recognise that, although replicated in the main elements of design and manufacture, both the main and auxiliary plant are subject to continued advances in materials, design techniques and in technology. Two years is considered to be too long to assume that there will be no changes in detail, particularly in metallurgy and welding, fasteners and fixings, data processing, numerical control, chemistry, corrosion, coatings, protection etc. The post-contract application of new developments, which may have been achieved by new sub-contractors, will have to be considered by the design-and-build contractors' staff and by the purchaser's Engineer, if approval is required by the contract, as the work proceeds through manufacture. Some changes when implemented will lead to detail changes in equipment, e.g. in size, the holding down bolts, the foundations, services, cables and control gear, and will need action by the Engineer to amend the detail of other service contracts. It is not possible to conceive that, in spending £200–300 million even in a replicated design, it will be possible to avoid alterations in detail completely, and project management must include a secure methodology to examine and contractually certify the amendments with jeopardising the programme or cost plan. This methodology will also include the need to deal with the rectification of the minor design and manufacturing errors which can arise. Some consultants put these minor rectifications at 70% of the total modifications needed in the post-contract stages (design contracts intended to eliminate post-contract alterations are a fallacy).

11.9 Variations

Fundamentally, the design-and-build contract assures that changes to the plant are the responsibility of the contractor and extra costs to the purchaser should not be involved, although the Engineer may have to consider extra time to complete. Even so, all contracts will have to include equitable methods for dealing with post-contract changes without disruption of the overall programme or leading to excessive extra costs. Fig. 11.2 sets out in principle the process for considering and authorising variations within the main plant contracts. The process for the service contracts is similar except that the amendments will be the result of a direct instruction from the Engineer to suit the alteration in the primary plant, usually affecting a simple change in size or extent of supply and involving a change in the contract price. It is stressed that only changes within the contract should be considered — amendments to the contract are not within the project manager's or Engineer's competence and must concern the purchaser before implementation. These must not be considered lightly or as a routine at working level — certainly no 'field' changes. Fig. 11.3 gives the principal delegations of the Engineer's powers and functions.

Amongst the recommendations of the Wilson Committee[3] was the suggestion that work would be better organised under the aegis of a few large 'super-contractors' using the least number of direct contracts and sub-contractors employing labour on site as possible; also to improve the financial incentives and reduce the provisions for risk particularly in a delayed completion situation, Cost-reimbursible contracts, with or without targets, could be used with advantage to both the purchaser and contractor (by securing the earliest completion). It is considered that this can only ensure that the price paid is maximised whilst risk to the contractor is a minimum. The CEGB's actual experience at Grain, Ince 'B' and Dinorwig has not borne out the aspirations of improvement, and subsequent practise has tended to revert back to 'small' manageable contracts on a definite work basis with the specialist manufacturers, including erection at a firm price with provision for CPA on materials and labour and re-measure where applicable, and this, in principle, is considered an essential in achieving success in the management of very large projects.

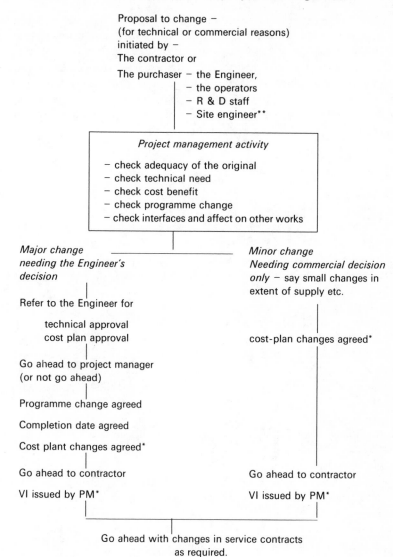

Proposal to change –
(for technical or commercial reasons)
initiated by –
The contractor or

The purchaser – the Engineer,
            – the operators
            – R & D staff
            – Site engineer**

Project management activity

– check adequacy of the original
– check technical need
– check cost benefit
– check programme change
– check interfaces and affect on other works

Major change   *Minor change*
needing the Engineer's   *Needing commercial decision*
decision   *only* – say small changes in
                                                    extent of supply etc.

Refer to the Engineer for

    technical approval
    cost plan approval   cost-plan changes agreed*

Go ahead to project manager
(or not go ahead)

Programme change agreed

Completion date agreed

Cost plant changes agreed*

Go ahead to contractor   Go ahead to contractor

VI issued by PM*   VI issued by PM*

Go ahead with changes in service contracts
as required.

*   Only if extra to the contract.
** No authorisation of field changes tolerated without reference back.

Fig. 11.2 *Procedure for issuing variations*

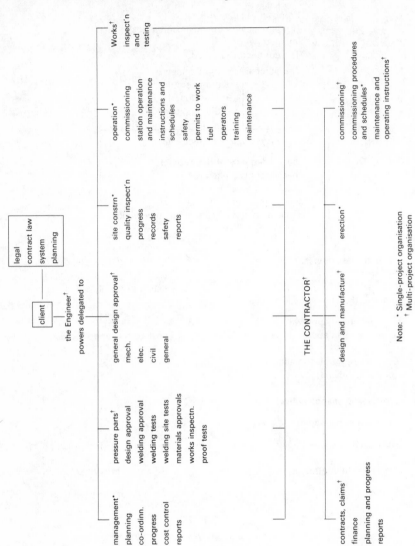

Fig. 11.3 *Responsibility and communications*

11.10 Systems designs and areas of work

Table 11.2 gives a list of the systems that have to be designed by the Engineer, whether directly employed by the purchaser, independent consultant, contractor or consortium of contractors. These system designs provide the basis for quantifying and specifying the works, and the details are built into the inquiries, tenders and contracts for the auxiliary plant and services whether purchased directly or through sub-contractors to the main contracts. Items 2, 3 and 4 are the subjects of substantial replication.

This list is followed in Table 11.3 which gives the headings for all likely areas of work to be carried out. The areas of work which would be replicated are also indicated. The rest would depend on actual site topography etc. and would be 'purpose designed' and contracted for. Some areas of work could be combined if so wished, to improve the contracting practice on site; e.g. coal and ash-plant foundations could be combined into a larger contract: even when so combined, the release of definite final design information, access to start work on site and completion is dependent on the needs of the master programme and the detail of the coal and ash-plant programmes. The only real advantage of large contracts appears to arise from the reduction in costs of site preliminary works; these, however, represent a small proportion of the total costs involved and approach insignificance.

The last point in this Section concerns competition. The CEGB and its predecessors have in the past been able to rely on competition in a wide and thriving industry to provide its generating plant and auxiliary works, and to establish the lowest price levels. The industry is now considerably reduced, with little or no competition to create the commercial pressure and entrepreneurial skills in a protected home market. The purchaser must be conversant with the realistic price levels in both the home and overseas markets, and if these are not realised with the limited competition or by negotiation, competitive price levels must be sought elsewhere. This was normal practice before nationalisation and has happened in almost every industry in the UK in the past decade or so.

Each generating unit is a highly capital-intensive project. The labour content represents less than 20% of the total costs. Direct client participation in the control of site labour or the execution of minor works should not be allowed to regulate or interfere with the major works, i.e. the main plant or structures, where responsibility to complete is firmly based on the contractors concerned. The main consideration of project management should be the completion of the project as a whole and not in the manipulation of minor works in a limited area with a negligible improvement, if any, in the overall time and cost.

11.11 Model conditions of contract

The simplest contract to purchase exists when a vendor agrees to sell, at a price, the goods or services sought by a prospective purchaser. In the case of

Table 11.2 *Areas of system designs*

1	Fuel, source, quality, availability, lifetime and basic economy
*2	Theoretical and practical steam cycle, optimisation and energy balance, vacuum and optimum recooled temperature. Final feed temperature
*3	Steam, water, make-up, blowdown etc. Boiler, turbogenerator, condenser, feed heating, extraction, feed-pumping, deaeration, make-up and rejection, condensate purity and rejection etc. Blowdown, heat and water recovery. Control, regulation, normal and emergency conditions, protection and procedures, fail safe features. Fuel firing systems. All leading to the main steam and feedwater, make-up, blowdown, drainage pipework systems and schematics
*4	Main building layout, plant relationship, location of main elements, loading, spans, structures Man and materials circulation, access to plant for operation and maintenance, vehicle access Handling and built in facilities. Sampling and chemical control facilities Ventilation, heating, lighting, amenities, canteens Security Operation control, permit-to-work systems and implementation Space optimisation and least building costs
5	Site layout, location of main and ancillary buildings, foundation options and costs, cut-and-fill balances, site levels, drainage and pumping, flood protection during construction and operation Men and materials circulation for construction and operation, road layouts and widths, parking and amenities Location of all services, water supplies and foul drainage EHV system and connections, in-site routes and protection during construction Mode of development — working out with minimum interference with operation road and rail systems, access to buildings and stores, security, safety fire, emergency conditions and access Permanent fence lines and in-house maintenance
6	CW system, make-up and purge, levels and pumping head, location of intake and outfall, off-shore recirculation, filling, drainage, maintenance

Table 11.2 *Continued*

7	Water-treatment plants and low-pressure water services and storage
8	EHV interconnections, auxiliary electrical system, essential electrical auxiliary system and emergency generators, 'black' start system etc.
9	Control concept, manning schedules, maintenance and operation needs
10	Coal offloading and storage systems, bunker feed systems
11	Ash and dust plant systems, storage and offsite disposal, gas clean-up plant and considerations, chimney height etc.
12	Amenities and architectural treatment, ZVIs, landscape preservation
13	Oil storage, fire risks and fire prevention

* Substantial replication.

generating plant, the vendor is a highly skilled, competent designer and builder in an international and highly competitive field of advanced technology, contracting to provide a reliable machine of a million horse-power or more for a service lifetime in excess of 30 years. The proprietary rights to the design, including the technical advances, lie with the designer. The extent of the works is too large to quantify and specify in full detail, and the vendor's expertise and responsibility to provide a working machine is fundamental to the contract. There can be no suggestion of partnership[5] under the model forms of contract as it at present exists. The concept of a third party, that is the 'Engineer', arose when utilities wished to buy high-technology equipment without having the in-house technical capability in sufficient depth. To overcome this deficiency, the purchaser used the services of suitably qualified consulting engineers to advise, specify and prove the acceptability of the completed works. These consultants were, and still are, professional people who have created for themselves, in addition to technical competence, an ethic of technical impartiality such that their judgement can be accepted as fair to both the purchaser and vendor. The Engineer was not a principal within the contract, which was between the vendor and purchaser, and the Engineer's involvement was only as far as both parties agreed. Both principals could resort to arbitration and to law within the terms of the contract in the case of a severe dispute, i.e. when the Engineer's judgment was not accepted. It is considered that the concept of the three acting in partnership is quite erroneous, and can only act to the ultimate disadvantage of the purchaser who is the only source of finance. To consider risk sharing would require a new form of contract, basically cost-plus, which would eventually lead to direct design and build. It is thought that the purchaser does not have the in-house capability at present, even if this procedure was thought to be desirable.

Table 11.3 Headings of the definite works

Civil	Mechanical	Electrical
Site surveys	Site drainage pumps and pipework	Site supplies and services
Site investigations	CR Boilers, fans, mills, mountings, auxiliaries	BT telephones etc.
Hydrographic survey	R Chemical cleaning	CR Main EHV switchgear
Camp and amenities	CR Turbogenerators and auxiliaries	CR Generator transformers
Off-site roads	CR Gas turbines, auxiliaries	CR Station transformers
Site clearance	CR BFPs	Unit transformers and auxiliary transformers
Preliminary site works	CR Condensing plant	CR Main generator connections
R Storm and foul drains sewage works	CR FH and deaerators	Main cables
R Site water and electricity	CR HP pipes and valves	Auxilary cables
R Temporary sidings	CR Blowdown, waste and recovery plant	Auxiliary switchgear and starters
R Fire protection and security	LP pipework and tanks	R Control equipment
R Piling	LP pumps	R Interlocking and sequence equipment
Main foundations	WT plant and chemical stores and handling	R Communication equipment
R Steelwork	Drain tanks and pumps	R Metering
R RC structures	Oil tanks, purifiers	Heating and lighting
R Superstructure	R Ash and dust plant	Control cables
Coal plant foundations	Coal plant	Aircraft lighting
Sidings, drainage	R Fuel oil tanks and plant	Sidings lighting
R Water reservoirs	Trackside equipment	Intruder protection
Water mains		R Computers
Off-site ash disposal		Construction-supplies
Lagoons, pipelines		Cable racks
R Cooling towers		Radiation monitoring
Off-shore works		

Table 11.3 *Continued*

	MUWPH and pipework	R	Jetty offloading equipment	Instrument workshops
	CW tunnels and pipes	R	Fire protection	Electrical workshops
	Screen chambers		Workshop and laboratory equipment	
	CWPH		Heating and ventilation	
	Service water PH		Stores equipment	
	Fire PH		Oil separators	
	Jetties		CW pumps, valves screens, strainers	
	Chimney foundations	R	Chlorination plant	
R	Chimneys	R	Hydrogen plant	
	Ash-plant foundations	R	CO_2 plant and storage	
	Storage tank foundations		Mobile coal plant	
	Switch houses		Vacuum cleaning plant	
	Amenity buildings		Weighbridges	
	Stores		Control air compressors	
	Ancillary plant buildings		Service air compressors	
	Fencing and security			
	Site clearance			
	Landscaping			
	Off-site landscape			
	Roads and carparks			

CR: Replicated and called forward from ongoing bulk contracts and updated for each project
R: Replicated design but re-tendered for each project
All others: Unique designs tendered for each project.

The purchaser, however, does have an in-house capability to provide the technical and contractual judgments of the Engineer, except the claim to impartiality, and there is a case for the consideration of direct purchase contracts without the involvement of 'the Engineer'. This is outside the scope of this Chapter, in which context, project management concerns the supervision of contractors and depends for its success almost entirely on the creation of a viable contract strategy and the correct interpretation and implementation of the conditions of contract that exist, and particularly the functions of the purchaser, the Engineer, the design-and-build contractor and the service contractors. These aspects are considered in further detail.

11.11.1 Owner, client, purchaser

Synonymous, and usually used indiscriminately, in the case of power stations this means the executive of the Board itself. Their statutory duty is clearly laid down – to provide electricity throughout England and Wales, to operate economically and efficiently, to preserve the amenity etc., and, to comply, the owner has to determine the future needs and to purchase any future plant that may be required. To do this, the owner has first to decide what fuels will be available, what plant he needs to meet both growth and replacement, and the price he can pay. He can be advised in-house or externally of the technology and the plant that is available. The manufacturing industries both at home and overseas are only too anxious to demonstrate their products and capabilites. The normal commercial case calls for the purchase by means of normal commercial contracts and procedures of proven plant already in operation somewhere, possibly with a slight extension of current technology indicative of the contractor's research and 'plus' products.

In the implementation of the current 'model' conditions of contract, the purchaser's functions are:

(i) To define the requirements in terms of fuel, output, modes of operation, loading rates, the technical parameters of the Grid system, fault levels etc., the availability, quality and quantity of water available for both boiler feed make-up and circulating water for the condensers, the completion dates, the methods of payment and the penalties in case of default on either side.

(ii) To define the economic parameters that will apply to the operation of the system throughout the lifetime of the plant – this will enable the design-and-build contractors to optimise their offers to the best advantage of the client.

(iii) To provide the site, the services and amenities, the access to the site, electricity, water, sewage disposal, flood protection, storm drainage, fencing, security (or not) etc.

(iv) To seek and obtain for the permanent works all the necessary consents, planning permissions, licences and approvals from all statutory and voluntary competent bodies including ministerial approval.

(v) To estimate the cost of the works and the incidence of expenditure, to provide all necessary finance to purchase the works including escalation and IDC, and to make the interim and final payments within the terms of the contracts.

(vi) To define any stores standards, interchangability, replication or any other definite preferences or options.

(vii) To define the additional components to be provided currently with the manufacture, as spares for the permanent works.

(viii) To provide operators, training facilities, fuel for setting to work and commissioning tests, water, electricity for auxiliaries and load for setting to work and testing.

(ix) To provide and maintain (or not) common amenities on site for the use of all contractors' personnel during the works.

The purchaser may use in-house staff or employ consultants and/or contractors to provide the above information and/or services in part or *in toto*.

11.11.2 *The Engineer*

The Engineer's function is to provide the technical expertise and to make judgments on behalf of the purchaser. probably the most important aspect is to ensure that the contracts are definitive, equitable and substainable in law, and do not lead to disagreement or conflict and delay in the execution of the works. It is essential to ensure that the inquiry specifications are clear and complete, and then to ensure that the vendor's offers fully comply or that any departures from the specification are identified, agreed in price and are compatible in all respects, practically and technically. The contract is for the supply of a complete working plant, and a high degree of skill is necessary to define the work to enable the vendor to prepare a competitive but equitable price, which is the only real basis for a contract. The normal functions are:

(i) If required by the purchaser, the Engineer may advise initially on all aspects of the feasibility and economics of the proposed development, the extent of the works to be included in each contract, the conditions of contract normal to the industry and the type of contract most suitable to the current commercial/industrial climate, the estimation of final costs, the preferred time scale and the overall plan. The Engineer for this stage of develop ment employed by the purchaser need not be the same person or group responsible for the construction and/or commissioning phases of the project; in fact, the use of people with skills in each stage is preferred.

(ii) Specify the extent of work required in each contract, the terminal points, the dates required to exchange final data and provide the definite physical detail of the plant to be supplied, access, intermediate and final completion dates, the relationship with other plants and contractors.

(iii) Set out the design practice and standards required, materials and manufacturing standards, set out the information required to be submitted for the Engineer's approval during the final design and manufacturing stages.

(iv) Appraise the tenders, identify any non-compliances, make recommendations, agree the extent of work offered, evaluate the contract price, prepare papers for the purchaser's formal acceptance etc.

(v) Agree the contractor's final proposals, detail drawings, schematics, programmes and schedules, and certify as necessary.

(vi) Inspect and certify materials, sub-contractors, manufactures, built-in components, works tests.

(vii) Inspect and certify the quality of the work on site before being covered.

(viii) Issue variations to the contract, including extensions of time (all within the terms of the contract — the Engineer has no discretion outside the contract).

(xi) Evaluate the work done and certify payments and define the completed work taken over. Witness tests on completion and issue final certificates.

(x) Monitor, record and confirm that the rate of progress to meet the operation date is being achieved or advise on action necessary.

(xi) Issue all formal instructions to the contractors (within the contract).

All these functions do not have to be carried out by one group or technical branch. In fact, this is not possible (or preferred); e.g. the approval of the design and fabrication of the presure parts is the responsibility of a recognised 'competent' authority, as are reservoirs, coastal works; overseas materials and work tests etc., all have to be delegated. It is important that the delegation and procedures for giving the Engineer's approval is set out in the contract documentation. The concept that all this technology should be channelled through one co-ordinator capable of giving judgment in all areas is not really practical or successful — non-expert approval has no value whatsoever.

It is considered that the Engineer's responsibility to assess and authorise any proposed change to a machine, or plant or structure, taking into account the effects on the programme and financial commitment, should be delegated directly to the responsible specialists concerned. The project manager's function is to co-ordinate, monitor progress against the 'plan', identify and clear bottlenecks both in-house and with the contractors, monitor expenditure and commitments against the budget; but under no circumstances should he be able to interfere without expert confirmation with the integrity of a million horsepower machine of advanced design in design, manufacture or operation. The same applies to less dramatic areas of plant and services, and the relationships are set out in Fig. 11.3. The project manager's duty is to monitor any significant changes, adjust the programme overall, allocate the responsibility, and call for the purchaser's approval if necessary.

Any area of engineering design, system design and contract supervision can be delegated to an external consultant if the expertise is so required by the purchaser; similarly to an Architect Engineer, which is a multi-disciplined organisation capable of the design of all the main and auxiliary systems in principle and, in some areas, the work in detail, but principally able to interpret and implement the purchaser's wishes and preferences, and carry out the day-to-day management of the whole enterprise. Particular manufacturers may be associated with an established architect engineer (or they may group themselves into a consortium and act directly) in which cases the competition for supply of components may be limited, but the overall 'package' may, through improved co-ordination, fewer interfaces, reduced IDC etc., if adequately financially guaranteed, be still financially attractive to the purchaser. If any of these agencies are employed, the purchaser will still need a number of experienced people in-house in an advisory function and to record progress and expenditure. With direct engineering costs now reaching 5–6%, the maintenance of a large in-house bureaucracy gets less attractive and the choice of management needs careful consideration. There is no case for the duplication of both directly employed staff and a consortium, irrespective of how described, to fulfil the Engineer's functions — duplication is just an added expense without added value.

11.11.3 Design-and-build contractor's function
The plant supplier is responsible for the design, manufacture and, in most cases in the home market, the erection of a plant which is guaranteed will meet the purchaser's specified output, performance, availability and lifetime. Some areas of the design are the manufacturer's protected patents, and there is a high degree of experience, research and development forming the 'plus' that the contractor is able to offer. It is this experience and entrepreneurial skill that is the basis of the contractor's judgements, and the price to be paid for this expertise. The only real and effective control of price available to the purchaser is by open competition between several competent suppliers. The only exception possibly is where the purchaser has included specific requirements for development outside the contractor's offer, which may call for the purchaser to accept some or all the risk for the implementation of his preferences. However, for all plant in the future replanting programmes, it is envisaged that the norm will be replication of plant already proved in service, and there is no case for risk sharing or extra expense to the purchaser. (After building $47 \times 500\,\text{MW}$ and $19 \times 660\,\text{MW}$ units uncertainty is incredible). The real difficulty when the competition is restricted is the ability to establish an equitable contract price.

11.11.4 Project manager's functions
The project manager's functions are:

(i) Collate the management, engineering and commercial views on the commercial strategy and set out in detail for the purchaser's formal agreement.

(ii) Set out the overall work plan to meet the operation date based on the shortest feasible completion at the rates normal to the particular industries involved, and without built-in unspecified time contingency.

(iii) Set out the practical detailed programmes for the system designs, the station design, the contract placing programme, each separate contract programme, the overall site construction programme and the commissioning programme. In addition to the networks and bar charts, all key dates are scheduled and included in the various contract commitments – these are easier to monitor and communicate. The plan is practical throughout without built-in margins between intent and commitment to the purchaser, and in the final form includes the contractor's detail as set out in each accepted tender.

(iv) Co-ordinate the preparation of the inquiry documents for the main plant and civil works: get the tender lists from the Engineer approved by the purchaser: set out the delegation of the Engineer's powers and communication routes that will be implemented in the course of the contract.

(v) Issue all inquiry documents, receive tenders, collate all appraisals and the Engineer's recommendations, summarise and prepare recommendations for the placing of the contracts by the purchaser.

(vi) Collate all commitments to expenditure, determine the incidence of expenditure and evaluate the interim and final expenditure, including the forecast inflation, the authorised contingency, engineering costs and interest during construction.

(vii) Prepare all documentation seeking authority to proceed.

(viii) Receive and collate all internal and external progress reports and assessments, summarise and define the overall position, issue all necessary instructions within the contracts to clear any delays. Thoroughly monitor all aspects of the clearance of the design, access and construction, and update the practical workplans as and when necessary, and make all the necessary adjustments in detail.

(ix) Systematically evaluate the work done using quantity surveyors and specialists as necessary, and certify all payments within the contracts: monitor commitment and expenditure against the approved budgets.

(x) Receive and record all claims for payments outside the contracts, including all the circumstances: refer to the Engineer and specialist branches responsible for the judgments with respect to arbitration and/or the purchaser.

(xi) Exercise the purchaser's functions with respect to site access, site amenities, services etc., security, electricity and water supplies.

(xii) Arrange for the allocation of site staff to fulfil the function of the Engineer's Representative, record progress, site inspections and evaluations, liaison with the operators, safety records and inspections.

(xiii) Co-ordinate access and works overall, arrange systematic progress meetings with the Engineer, all contractors, operators, commissioning staff. Produce all minutes and contract records, action sheets, log books and site photographs.

(xiv) Issue monthly reports on the practical and financial situation, and advise the purchaser of any irrecoverable deterioration. Call for and issue construction status reports and design status reports from the Engineer as when thought to be necessary.

It should be stressed that the most important aspect of the investment is the operational performance and availability of the generating unit, i.e. the boiler and turbogenerator, and these are the direct contractural responsibility of the design-and-build contractor. It is the quality of the technical decisions by the contractor in the plant design that is most important, and the responsibility for the approval of the design lies within the specialist functional engineers. There has tended to be too much emphasis on site works and minor aspects of employment which are considered to have in the past obscured the real issues needing management action and control.

11.11.5 Site engineer
It has become fashionable to employ a site manager with a wide-ranging brief, but with very vague terms of reference and limited actual responsibility. In the concept of project management set out in this Chapter, the responsibility of site staff is limited to that accorded to the Engineer's Representative in the model conditions, and no further, namely:

(i) Record and approve the ground conditions before permanent work.

(ii) Examine, approve and record the acceptability of all work before being covered up or welded in. Arrange access for the 'competent' inspection of pressure parts and the like.

(iii) Record progress in all areas of work and report systematically on any deficiencies, recommend actions within the contracts but no site orders.

(iv) Witness all necessary proving tests before setting to work and record; issue the necessary completion and take-over certificates, safety clearances as required by the contracts.

(v) Carry out all the necessary valuations, certify the goods received and works taken over. Record and protect the purchaser's property as necessary.

In the interest of firm control, it is considered that, other than the rectification of any unsatisfactory work within the contract, the site staff is not authorised to issue any variation orders on the contracts calling for any field changes in the design or extent of supply (day works excepted) without any prior confirmation of the Engineer or the project manager in the case of programme or sequence of work. The responsibility for meeting the programme and getting the plant work lies entirely with the contractors involved. Similarly, the responsibility for industrial relations lies with the actual employer, i.e. the contractor, and conflict or diminution of the responsibility to complete must be avoided. The sharing of responsibility for delay can only result in extra costs to the purchaser — he is the only source of finance.

11.12 Concluding remarks

This Chapter has attempted to set out some of the key aspects in managing the large construction programme envisaged in the replanting of the existing coal-fired and AGR stations expected to commence at the turn of the century. The objective must be to secure the completion of each unit to both time and cost. It is essential that modern economic designs for the generating plant itself, both coal- and nuclear-fired, and all the auxiliaries and services are considered, and working plans prepared in detail over the next decade, especially including the suitability for effective and continued replication.

The possible advantages of simplifying the contract procedures by updating the functions of the purchaser and the design-and-build contractor, with the elimination of the Engineer in the model conditions should also be considered. All the main plant and service equipments will continue to be purchased on a design-and-build basis from competent contractors; and, without doubt, the most important aspect is to ensure that a viable contract strategy, based on competition under terms equitable to purchaser and contractor, is prepared and agreed, and an efficient, 'slim', non-bureaucratic project-management organisation is trained and motivated to secure economic results comparable with the best of the sister organisations overseas.

This Chapter expresses the views and opinions of the author and carries no suggestion of endorsement by the Central Electricity Generating Board or the Generation, Development & Construction Division. Nevertheless the paper would not have been possible without the experience gained by the author within the CEGB and the past assistance of the many colleagues in both the CEGB and the power-station construction industry. This support is gratefully acknowledged.

11.13 References

1 CEGB Annual Reports

2 A report on the operation by the Board of its system for the genration and supply of electricity in bulk. Monopolies and Mergers Commission report on the CEGB, May 1981

3 Report of the Committee of Enquiry into Delays in Commissioning CEGB Power Stations. Wilson Committee Report presented to Parliament, March 1969

4 'The costs of generating electricity in nuclear and coal fired power stations.' Nuclear Energy Agency − Organisation for Economic Co-Operation and Development (NEA-OECD). A report by an expert group 1983

5 BURBRIDGE, R. N. G.: 'Some art, some science and a lot of feedback' *IEE Proc.* 1984 **131**, Pt. A, (1)

6 MASDING, H.: 'Construction of power stations designed and built as single units' *I. Mech. E. Proc.* 1982, **193**, (36)

7 MASDING, H.: 'Design for overall economy − UK power stations' I. Mech. E, 1979, Paper C104/79

8 'A Review of some of the factors affecting the price of electricity' *Electrical Power Engineer*, March/April 1984, pp. 93–95

Chapter 12

Project management: A review

J. W. Currie, R. M. Gove and A. F. Pexton
South of Scotland Electricity Board

12.1 Introduction

The South of Scotland Electricity Board (SSEB) is an all-purpose utility responsible for providing electricity supplies to 1.6 million consumers. Since its
formation in the mid 1950s, the Board has undertaken a large number of major
power-station construction projects with associated developments to the Grid
system, and is currently engaged on construction of the 2 × 700 MW AGR
station at Torness. Table 12.1 summarises project performance on the major
coal, oil and nuclear projects undertaken by the Board.

This Chapter outlines some of the experiences encountered, practices
employed and developments introduced by the Board in the course of its
history. Naturally these developments have not been carried out in isolation, a
particular feature of the Electricity Supply Industry being the extent to which
ideas and experience are interchanged with the common aim of improving UK
power-plant project engineering practice. The structure of the contracting
industry has altered markedly over the past 30 years, and some of the more
significant changes are outlined.

Success in any major project calls for continuous dedication by the client from
the outset to the end of plant life, and this quality has been available in full
measure on all SSEB projects. This dedication is best complemented by building
systematically on previous experience, given a reasonable amount of replication
of design.

12.2 Client engineering

Even in a major engineering-based organisation individual projects can make
excessive demands on in-house resources throughout the design, construction
and commissioning periods, and on the majority of the early Scottish stations,
as well as at Kincardine, Hunterston A, Cockenzie and Longannet, Consultants
were employed for overall engineering design and site supervision, and were

Table 12.1 *Major SSEB power station projects*

Station	Size of units	Type	Start of construction	Date of synchronising first unit		Cost overrun, %
				Planned	Actual	
Hunterson A	2 × 150	nuclear (Magnox)	1957	1961	1964	75
Kincardine HP	2 × 200	coal	1959	1962	1962	Nil
Methil	2 × 30	slurry	1962	1965	1965	7
Cockenzie	4 × 300	coal	1962	1966	1967	0.4
Longannet	4 × 600	coal	1964	1968	1970	12
Hunterson B	2 × 660	nuclear (AGR)	1968	1973	1976	16
Inverkip	3 × 660	oil	1970	1975	1976	10
Torness	2 × 700	nuclear (AGR)	1980	1986	–	

overseen by small client project management teams. Contracts co-ordinated by the Consultants were placed following competitive tenders from established manufactures or, in the case of Hunterston A, from a nuclear consortium. In the case of the fossil-fired plants, these arrangements proved satisfactory for their time. However, with growing dependence on the reliable operation of a small number of large and technically complex generating units, the need for a robust in-house project/technical resource capable of solving problems arising in the design construction and operation phases, and of ensuring continuity and feedback of operating experience, became increasingly clear.

The lesson was brought home forcibly to SSEB in the early 1960s in their first involvement with nuclear power on the construction of Hunterston A. The lack of in-house nuclear expertise, compounded by the general inexperience of contractors in nuclear technology at that time, made it impossible to provide the necessary project control and ensure corrective measures were taken to avoid completion delays and cost overrun. The Board's Hunterston A experience highlighted the need to strengthen the engineering resource base and the lesson was well learnt.

12.2.1 Project support

A prime aim in strengthening the Board's ability to handle large projects has been the improvement of support for project and operation teams under the various engineering and administrative disciplines. Some new groups were formed in the mid 1960s, but most disciplines were already covered and the task was mainly one of strengthening existing units. Fig. 12.1 outlines the main support groups.

In all areas, but particularly in the technical field − design, technical services, nuclear safety − the time scale to build up a fully effective resource can be long. At the formative stage enthusiasm can overcome a great deal, but the establishment of experienced and professional units probably takes as long as ten years, and perhaps still more to settle down with effective working relationships.

In recognition of the growing importance of specialist advice, a number of the groups were progressively strengthened through the 1970s, and to some degree that development is still continuing. For example, the control and instrumentation group has been expanded in recent years. Although control and instrumentation technology generally costs les than 5% of the capital value of a power-station project, this equipment can cause extensive loss of availability of expensive plant. So rapid has been the rate of development in control and instrumentation that it is often difficult to obtain spares to maintain relatively modern installations, and in many cases it has been preferable to replace with updated equipment. In recognition of these problems the control and instrumentation group in SSEB therefore not only assess and recommend the type of equipment to be purchased, but also carry responsibility for overseeing its setting to work, both on operating stations and on new projects. This policy is paying handsome dividends in providing continuity throughout the Board's installations. This approach — of maintaining a skilled specialist resource to cater for particular aspects of the work — is not new, and has indeed been followed for many years in other parts of the Board's Engineering Department, such as protection and telecommunications.

Fig. 12.1 *Support groups*

12.2.2 Project team

With the emphasis on disciplines, optimisation of the size of a dedicated team to manage a major project needs careful judgement. The team must be strong enough to control and direct the full range of contractors, but sufficiently compact to avoid the temptation to take on tasks which rightly belong to the contractor or to SSEB specialist support units. For Torness, with a capital cost well in excess of £1billion, the SSEB project team consists of 20 engineers, supported by a site team of 30. Some of the project engineers have worked with the support groups, and some will return there — though probably to a different group to broaden their experience — at the end of the project. This is now a well established pattern, although it clearly depends upon the continuity available on project work.

Meanwhile the overwhelming need throughout the whole period of a project is to bring to bear the relevant expertise, determination and staying power in the project team. The right leadership is critical, and the importance of a few key individuals on a large project should never be underestimated.

12.3 Structure of the nuclear industry

Nuclear consortia have played a major role in the engineering of SSEB's three most recent major power station projects, Hunterston B, Inverkip and Torness.

In 1965 it was decided that, rather than employ overall consultants, SSEB with its strengthened support units would directly manage a comprehensive contract for Hunterson 'B' (Fig. 12.2). Tenders were invited from the three nuclear consortia then in business. They submitted three very different designs of AGR stations and The Nuclear Power Group (TNPG) were successful.

Construction of the next plant, an oil-fired station at Inverkip, started in 1971. This was a time of particularly heavy project activity in SSEB with both Hunterston B and Longannet making heavy demands on resources, and Cockenzie still encountering significant early operational problems. To control the Inverkip plant contracts directly would have required a large increase in design and project resources which would have been difficult to absorb subsequently. It was therefore considered that, with SSEB's experience in the design problems of fossil-fired stations, the benefits available from deployment of the project management skills of the nuclear consortia would outweigh any shortfall in their experience on oil-fired plant, and comprehensive tenders for Inverkip were invited from the two nuclear consortia which then remained. Again TNPG were successful. they had acquired their skills on nuclear power-station contracts over a period of 20 years, and had a good track record of successful project management. In the event they were to find little difficulty in adapting to an oil-fired station.

The structure of the nuclear consortia was such that selection of the major member companies as contractors was almost automatic, and competition was restricted to the smaller contracts once the overall bid had been accepted. This approach yielded the benefit of continuity of experience and working relationships, but limited the extent to which the Board could apply the full rigour of competitive tendering to the individual main plant items.

It is relevant to note at this stage that the North of Scotland Hydro-Electric Board's (NSHEB) 2 × 660 MW oil-fired station at Peterhead has an important place in the recent history of Scottish power-station construction. Site work started in 1973, and the project was divided into five major packages controlled by managing contractors selected by competitive tender. Each was a major plant supplier but also had a co-ordinating role over a number of supporting contracts, some of which were placed directly by NSHEB and handed over to the co-ordinating contractor in an agreed manner. The station was built within its original budget, and success was in no small measure due to the response of the contractors to the greater reliance placed upon them. Some services such as specialised design support and monitoring of manufacturing progress were supplied by SSEB, but the project demonstrated very effectively how a small project team, given good support from its contractors, could control a project of significant scale.

Fig. 12.2 *Hunterston B Power Station reactor charge hall*

In 1975 the two then remaining nuclear consortia were combined to form the National Nuclear Corporation (NNC). Along with the United Kingdom Atomic Energy Authority (UKAEA) and the General Electric Company (GEC), member companies from the two remaining consortia jointly became shareholders, and the General Electric Company were given a supervisory management contract. Hence national expertise on nuclear-power-station design and construction was brought together in a single organisation. The shareholding in NNC of different companies with similar product ranges provided the opportunity to encourage competitive tendering for each main plant area. Also, to optimise the deployment of skills within the industry, an understanding was reached with the CEGB that NNC would in future concentrate on the 'nuclear island', whilst other major contracts for the 'balance of plant' would be managed directly by the Generating Boards. It was on this division of work that Torness (Fig. 12.3) was launched, to common procedures developed jointly with CEGB for their twin station of Heysham II.

Fig. 12.3 *Torness construction site, 1985*

NNC were made responsible for preparing a common outline station design for Heysham and Torness and for the detailed design of the reactor islands and civil works. Individual component design was carried out by the plant contractors except for some specialised items which NNC dealt with themselves. The

Generating Boards exercise overall project control and directly manage contracts for turbine generators, cooling water plant, diesel generators, heavy electrical equipment and other 'balance of plant' contracts. The 400 kV and 132 kV switching stations at Torness and the transmission connections also form separate contracts directly handled by SSEB (Fig. 12.4). NNC acts as project manager for deisgn, manufacture and construction of the nuclear islands. At Torness NNC were also given responsibility for managing some interface contracts which extend beyond the nuclear island. These include civil works, cabling and much of the pipework.

Fig. 12.4   *400 kV metalclad switchgear within substation building at Torness*

From the SSEB viewpoint these arrangements combined the best features of Hunterstone-B/Inverkip and Peterhead, and optimised developement of the experienced executive officers and specialised staff of the nuclear consortia. It was in line with Government policy at the time of establishing NNC as a strong company and the national centre of expertise in nuclear-island design and construction — formalised most recently in July 1981 in Cmnd. 8317.

NNC shareholders were not, however, prepared to allow the company to offer joint and several guarantees nor accept levels of contractural risk in keeping with the scale of the contracts they would be managing, and this prevented them from being recognised with full status as main contractors on Heysham and Torness. Their role has therefore become one of Agents for the Boards. With the Boards shouldering virtually all the financial risk, they played a major role in framing the procedures, methods and conditions applied to the placing of contracts to be managed by NNC.

In practice on Torness, any shortcomings have been outweighed by the effective and co-operative working relationships which have been developed between client and agent managers and staff deployed on the project. There can be little doubt, however, that the agency role has not strengthened NNC's general standing, although SSEB were hopeful in 1980 that this would be resolved in the future as the new company grew in stature and confidence, and shareholders were prepared to increase capitalisation.

Despite these shortcomings, the working structure at Torness is still a good model on which to build for the future, not only for projects in the UK but to put NNC in a position to seek export orders if they are prepared to take a stronger contractural role.

12.4 Establishment of a project

Once the need for a project is foreseen, there is an extensive pre-construction period. With the increasing demands of planning authorities and amenity bodies and growing pressure from the public in general, it is becoming necessary to deploy considerable resources to prepare comprehensive submissions for planning and consultative purposes, in addition to the extensive effort required on detailed design. This puts heavy demand on the client's technical resources to ensure that he can establish the detailed parameters, features, environmental impact and safety characteristics of a project ahead of the planning procedure. In certain circumstances where planning procedures are of extended duration, this can lead to the requirement for detailed specifications to be prepared and tenders sought before planning permission is obtained, with the risk of much abortive work.

12.4.1 Public inquiries

It is beginning to appear improbable that any future power-station site in the UK will be opened without a prolonged public inquiry. The time taken by inquiries into SSEB projects is listed in Table 12.2

The recent trend of increasing public pressure is exemplified by Torness and its associated projects. It is worth noting that, after an eight-day inquiry in 1974 for an ultimate installation of up to 5280 MW on the 350 acre site, an inquiry in 1982 into the proposed routes for the 400 kV transmission lines to connect the station to the grid took 22 days, although there had been full consultation and agreement earlier with the various planning and consultative bodies. The recommendation of the reporter at that inquiry was that an alternative route be sought over a large part of one of the two routes proposed by SSEB. A subsequent application, based on the alternative route preferred by the inspector, then met objection from one of the local authorities, and in 1984 a second inquiry into this alternative lasted a further five days. Also in 1984 there was an inquiry into a proposed one-acre railway siding site 1 km distant from Torness for dispatch of irradiated field. This lasted 10 days. It is indicative of the trend that an inquiry into a small railway siding in 1984 should last longer than the main station inquiry in 1974. These periods may seem relatively trivial compared with the CEGB's mammoth 340 days at Sizewell, but the trends are notable.

Table 12.2 *SSEB Public inquiries for engineering installations*

	Year	Period, working days
Power stations		
Hunterston A	1957	13
Inverkip	1969	6
Carriden (not constructed)	1974	7
Torness	1974	8
Transmission lines		
Dalry/Kilbarchan, 400 kV	1970	2
Westfield/Abernethy, 275 kV	1972	2
Inverkip connection, 400 kV	1973	3
Neilston/Kilmarnock, 400 kV	1975	2
Torness connection, 400 kV	1982	22
Torness connection (amendment)	1984	5
Rail siding		
Skateraw (Torness)	1984	10

The manpower demands of such public inquiries have become substantial. For example, taking all factors into account, it is estimated that to prepare and present Board's case at the Torness transmission-lines inquiries absorbed in total some 800 man-weeks of senior engineers' and managers' time.

Each inquiry brings its own trauma, and the approach to be adopted is not amenable to generalisation. An inquiry team ranging over the various disciplines has to be identified in support of individuals selected to give evidence. A common theme has always been the extraordinary team spirit generated among those involved in promoting a proposal.

12.4.2 Need for change in public inquiry procedures
The need to change public inquiry procedures, particulary in the light of Sizewell, is much discussed but there is a dearth of directives. The difficulty is to find an acceptable alternative which ensures the proposal is sufficiently exposed and examined, allows all citizens really affected to express their objections, but does not involve a full local re-examination of national policy issues on each and every occasion. Certainly attention must be given to streamlining the evaluation

of major high-technology projects, which, as evidenced at Sizewell, are now being debated in public to the finest detail. Whilst a Reporter may start with the objective of controlling the range of coverage to be allowed, experience is showing that, without an understanding on Government policy in advance, it is proving extremely difficult to practice to apply restraint.

With the Council of Engineering Institutions (CEI) now providing a focus for the engineering profession, one possibility might be for CEI, working through the major professional engineering Institutions, to be invited to appoint expert independent panels to make recommendations on technical issues of national importance. The information from such investigations, if it were endorsed as Government policy after Parliamentary debate, could then be presented to an inquiry into local issues, so enabling the inspector to focus and limit the range of detailed debate at the inquiry. Such an approach would need much development before it could gain public and statutory support, but, rather than a full two-stage inquiry, it might offer scope for more efficient utilisation of resources and increased acceptance of decisions taken in the overall public interest.

12.4.3 Relations with local authorities and amenity bodies
Once planning permission has been obtained, there is a commitment on the promoter to ensure that all agreements with local authorities and amenity bodies are rigidly enforced. The SSEB attach great importance to meetings such commitments. This demands meticulous attention to detail throughout construction and into the operating phase.

Local employment, always an important theme, has acquired particular significance in recent years. At Torness, SSEB encourage recruitment of construction labour locally as far as suitable skills permit. The labour force, including staff, built up to 1000 in the second half of 1980, to 2000 in mid-1981, to 3000 at the beginning of 1982, to 4000 in mid-1984, and peaked in mid-1985 at 5900. Apart from a period in 1981/82, the proportion of recruits from within 25 miles of the site has been held above 40%. The percentage from Scotland has been consistently around the 70% mark. Over the period, some 500 youths aged eighteen and under have been taken on by contractors, the majority undergoing industry training with the prospect of subsequent employment.

As a new venture at the start of Torness, to provide a focus for working relationships with the five local authorities with interest in the station construction, a local-authority project-liason committee was established, and this meets about four times a year. It has been a useful vehicle for information interchange and it could form a pattern for the future. A similar approach has subsequently been adopted elsewhere in the Board's district for smaller projects of more localised interest, such as extensions to land reclamation at the Longannet ash lagoons.

12.4.4 Appointment of main contractors
A project can be regarded as established when statutory consents and financial approvals have been obtained, terms agreed with local authorities and amenity

bodies, and the main contractors have been appointed. The single most important ingredient of any project is the formation of a force of highly competent well co-ordinated contractors, each with a clear-cut role and appropriately drawn interfaces with other contractors.

It is only in special circumstances that SSEB do not adopt competitive tendering, and reference has been made to the constraints in recent years inherent in the nature of the nuclear consortium system. Some single tender action has been carried over to Heysham 2/Torness. With design firmly based on Hinkley/Hunterston, as the successful AGRs, it was decided jointly by CEGB and SSEB that a number of contracts for the more specialised nuclear plant should be negotiated with manufacturers who had been engaged on Hinkley and Hunterston. In all, nine major contracts and two sub-contracts were placed through negotiation with nominated contractors. These include the steelwork inside the pressure vessel, boilers, gas circulators and refuelling equipment. As part of the negotiation of these contracts, dedicated manufacturing facilities were financed to overcome some of the difficulties encountered on the earlier AGRs where existing workshops were adapted on an *ad hoc* basis. These new facilities have been an important factor in the timely delivery of the major Heysham 2/Torness components.

A condition of tendering for the negotiated contracts was that the make-up of the price should be fully disclosed, with supporting information as required. The compilation of tenders revealed manufacturing hours and rates, overheads, material procurement, capital plant, site-erection labour costs, contingencies and commercial provisions. With the co-operation of the selected contractors an assessment panel consisting of engineers, contract officers and a management accountant examined the tenders in great detail. Where appropriate, the panel recommended adjustments to the tender price. The major areas for negotiation were risk assessment, contingency, overheads and profit margin. The adoption of negotiated contracts for certain components at Heysham 2/Torness does not, however, preclude the re-introduction of competitive tendering for AGR components in the future, although changing contractors could lead to detailed component differences due to the pre-existing design rights of the present contractors.

There has been significant rationalisation of the UK power-plant manufacturing industry in recent years. In the early 1950s bids could be sought from up to six boilermakers and five turbine manufacturers. For much of the major plant on AGR and fossil-fired stations, including boilers and turbine generators, the number of possible UK tenderers is now two. There are still a small number of areas, such as AGR refuelling equipment, for which bids could be sought from up to four competent manufacturers. If a structure similar to the present one can be sustained against a reasonable plant-ordering programme it would seem to provide not only the route to competitive UK project costs, but also good long-term prospects for improved efficiency and success in the export field.

12.5 Project management

By the time the civil and main plant contractors have been appointed, formation of the project team must be well advanced. A shortcoming of early large power-station projects was that, owing to the urgency of meeting high load growth, insufficient time and effort were devoted to pre-construction work. The difficulties of moving from plant in the 60/120 MW range to the 500/600 MW range was underestimated and this brought technical problems with the wide range of designs then available, and also origanisational problems with the increased scale of the site work force.

Hunterston B, for example, was a prototype and encountered a range of problems during construction and comissioning. Many of these were due to the fact that prototype plant was being installed, and some of the problems would not have been foreseen with any amount of pre-planning. Others, however, could have been identified by taking design and development to a more advanced stage before construction. At Inverkip and Torness, the two most recent SSEB projects, provision was therefore made for more extensive preparatory work before the start on site.

Fig. 12.5 *Inverkip Power Station turbine hall*

12.5.1 Design contract
The first SSEB design contract was placed with TNPG when they were awarded the order for Inverkip (Fig. 12.5). This was the first large SSEB oil-fired station, and the specification required it to operate as rapid start-up and load following plant. Changes introduced during the course of the design contract, which lasted

12 months, included an increase of boiler-drum diameter, re-design of drum internals, increased furnace circulation ratios, change of boiler-tube material, and a major revision of the turbine-hall layout to improve maintenance access and lay-down space.

A similar concept was followed for Torness and CEGB's Heysham 2 station, where the pre-construction design period lasted for approaching two years. NNC in conjunction with CEGB and SSEB staff developed the overall station layout and design for both sites, with agreed nuclear-island plant studies carried out by the manufacturers to establish a common nuclear-island design for these and future UK AGR stations. After completion, the Boards placed separate design contracts directly with the turbine generator manufactures during the same period. Progress was also made on definition of quality-assurance plans and detailed plant programmes. Job-inquiry specifications were prepared for ancillary plant items, and fault studies were carried out to enable completion of the pre-construction safety report required for licensing.

Despite the principle of replicating Hinkley/Hunterston, the design of Heysham 2/Torness introduced detailed engineering developments not only to incorporate improvements indicated by operating experience, but even more significantly to meed updated safety standards for seismic criteria and segregation of emergency supplies. Fig. 12.6 compares the Hunterston and Torness pressure vessels and the components inside. Torness has a slightly bigger core to give improved performance and a larger pressure vessel to give improved maintenance access. Thus, although great benefit undoubtedly accrued from experience at Hunterston B, a fresh start had to be made on the drawing board.

For a future AGR, even though the generated output would be increased, the bulk of the Heysham 2/Torness components would be replicated and many of the half-million drawings could be re-used. The new station could therefore be taken to a highly advanced stage of detailed design in the course of the pre-construction design period.

12.5.2 Main construction period

Textbooks have been written on how to handle projects once construction has started on site. In the end success comes from an amalgam of experience, clearly defined philosophy, hard work and determined management. Every day brings its own problems to be promptly dealt with and every stage of the work is crucial to success.

Steady progress is, however, being made generally on improving quality control, and on refining the structure and professionalism of site and contract management. There is merit in further attention to contract terms and their thoughtful implementation, but real success can only be achieved by the continued evolution of strong project and site teams within the contractors' organisations. For this to happen construction engineering must be seen as a rewarding profession in a stable industrial infrastructure, with systematic commitment to an on-going programme of work.

In recent years major strides have been made by UK contractors in completing the assembly of larger work packages in the factory. This is demonstrated not only in the AGR facilities referred to above, but in a wide range of plant for AGR and fossil-fired stations and for transmission installations. The Torness turbine-generator manufacturer has, for example, delivered direct by sea a number of large factory-built modules. These include complete high-pressure and intermediate-pressure turbines, tubed condenser sections (Fig. 12.7), lubricating-oil and fire-resistant fluid-system packages, and intermediate-pressure steam chests complete with loop pipes. The heaviest load delivered to Torness was the reactor gas baffle which weighed 1300 tons on its temporary support structure.

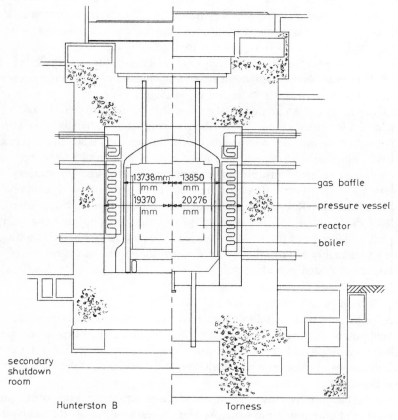

Fig. 12.6 *Cross-sections showing comparisons of reactor pressure vessels*

There is perhaps still a tendency for some parts of the UK construction industry to under-estimate the extent of planning required, and in particular the need to complete detailed planning at an early stage. On recent SSEB sites the Board and NNC planning offices have been built into an important control point through which information is made available to contractors. A simple technique used to good effect has been the establishment at the outset of a formal six-monthly review at managing-director level of the work of each contractor. Manufacturing and site progress is brought into sharp focus in the contract conditions, with key dates and interruption to cash flow for failure to perform rathern than reliance on liquidated damages, which, because they are settled so far into the future, do not have immediate impact. Fig. 12.8 shows achievement against target on a number of the more important key dates for Torness.

Fig. 12.7 *Condenser section being delivered to Torness construction site*

Prior to Torness, invitations to tender for civil-engineering, electrical and mechanical work placed the obligation for insuring works, plant and third-party liability on individual contractors. For the construction of Torness, however, SSEB assumed responsibility for the material and liability insurances, with contractors providing cover for their own employees and plant and for plant being manufctured on their premises. This arrangement, known as 'project insurance', has resulted in savings in premium costs over the previous approach, together with improved cover and more effective management control on a day-to-day basis.

12.5.3 Industrial relations
Improved interchange of information has been introduced progressively in the industrial relations field. Day-to-day issues continue to be dealt with by site

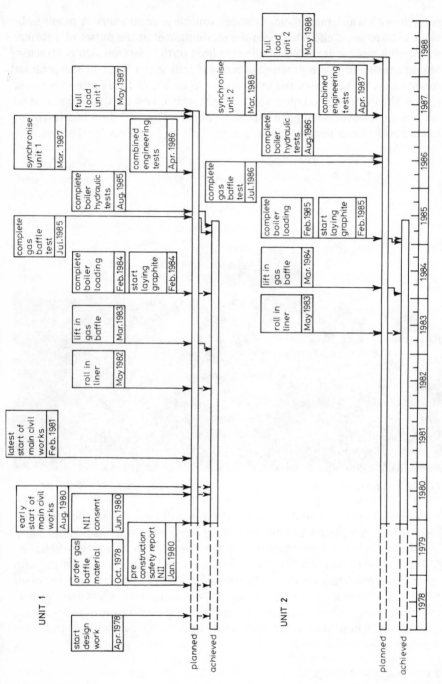

Fig. 12.8 *Torness construction programme showing comparison of actual and planned dates*

managers, and the need for greater professionalism in handling industrial relations on site is widely recognised. Many difficulties can, however, be avoided by the establishment of common industrial-relation policies across the site and joint consideration of problems when they do arise before action is taken.

Over the five years of construction so far at Torness, the time-lost figure is less than 1%, due in no small part to the disciplines for both contractors' management and labour brought about the National Agreement for the Engineering Construction Industry (NAECI). This came into force in September 1981 after years of endeavour to find common ground for a properly structured approach to site industrial relations by the clients, unions and contractors involved in both the heavy-engineering and the oil and petro-chemical sectors. Torness was one of the first projects to be 'nominated' under the agreement, and although almost all mechanical and electrical con-tractors are operating individual productivity bonus schemes to suit their own type of work, pay-packet compatibility has been successfully maintained. All operatives within the scope of NAECI have been employed on the basis of their willingness to undertake shift work if required and the general pattern has been double day shift for operations in the reactor pressure vessel, on the charge machine and on cabling and certain pipework activities with normal day shift elsewhere. When necessary, a small night shift has been sucessfully superimposed on these shift patterns to deal with specific problems. This flexi-bility of shift-working arrangements to match the changing requirements of individual contracts has broken new ground for SSEB at Torness. The Project Joint Council (PJC) set up under the NAECI arrangements has functioned well, with strong representation from all signatory unions in the form of area officers. The principal contractors have also combined well to provide continuity of employer-side representation from both senior head-office staff and site managers.

Considerable effort has been expended by the Board in vetting the industrial-relations arrangements and bonus schemes of individual contractors before they start work and in auditing them periodically. The independent audit team, staffed by the Board's management-services personnel, collects and issues to the PJC comprehensive information on numbers of in-scope contractors and employees, their recruitment area, turnover, weekly bonus payments and over-time. Despite earlier concerns, the provision of such extensive information, which is also available to shop stewards on the PJC, has proved nothing but beneficial.

The PJC has proved its worth in monitoring industrial-relations practices across the site and in dealing locally in a prompt and informed manner with almost all disputes and claims referred to it from company domestic level. It has also taken considerable interest in employee safety and welfare, and most important of all has maintained a positive and constructive interest in the project being kept up to programme.

12.5.4 Welfare

As well as recognition of the need to improve the standard of safety, working conditions and amenities on site, there has been a progressive move on Scottish power-station projects to provide attractive residential accommodation for construction workers where there is a limited pool of skilled labour locally. The standard was set initially at Longannet, and improved by NSHEB at Foyers (on a smaller scale) and Peterhead. The Torness construction village, 3 km from the site, embodies this joint experience. As well as providing 640 individual bedrooms, it offers restaurant facilities, shop, games hall, medical centre, laundry, film theatre, general recreational rooms, and a fully landscaped residential caravan park for families. Fig. 12.9 shows a plan of the village. The operation is largely self-financing apart from the initial capital cost, and is fully justifying the philosophy that good facilities enhance the standing of construction workers and promote good integration with the local community through the sharing of facilities.

12.5.5 Commissioning

Although station operating staff need to be in post for the main system tests, the commissioning task for a large power station is too complex and extensive to form part of a simple handover from construction to operating staff, both of whom have their own commitments throughout the commissioning period. Recent practice has been to form a commissioning team, which takes over individual plant items from the construction engineers and completes the combined engineering tests for handover to the station staff. At Longannet the commissioning team was formed from SSEB, consultants and contractors' staff and at the more recent stations joing NNC/SSEB teams have worked under NNC leadership. The overall process is under the direction of the station manager, who chairs the commissioning committee.

12.5.6 Plant modifications

The modification of plant to overcome problems encountered in commissioning and early operation, and to improve long-term plant reliability, is an on-going process at all major stations, which makes heavy demands on recourses. Hunterston B was commissioned in 1976 and SSEB still have a team of about ten construction engineers on site responsible for plant modifications and extensions. They have undertaken a wide range of work from substantial plant additions to detailed refinements of ancillary equipment. To varying degrees similar work is in progress at all SSEB stations, and the pattern will no doubt be followed at Torness, although to a lesser extent in view of Hunterston B experience having been built into the design. These project teams, along with the station staff and the design engineers who ensure continuity of technical standards, take credit for the progresive improvement in plant availability. Table 12.3 shows the availability of SSEB plants in recent years.

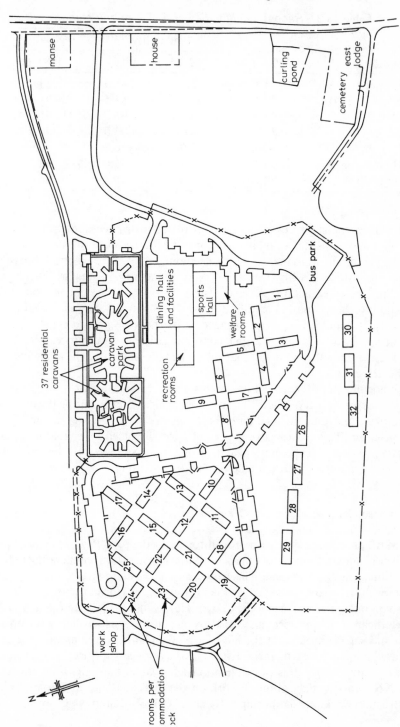

Fig. 12.9 *Torness construction village*

Table 12.3   *Recent availabilities of SSEB power stations*

Station	1981–82	1982–83	1983–84	1984–85
		Availability, %		
Hunterston A	69.3	72.1	81.9	83.6
Kincardine HP	70.7	48.1	61.5	91.0
Methil	80.0	73.3	86.0	91.0
Cockenzie	73.3	84.2	91.0	75.9
Longannet	66.9	83.3	69.8	69.9
Hunterston B	55.7	71.7	75.0	78.3
Inverkip	98.1	96.0	98.8	96.0

12.5.7 Feedback
The SSEB specialist groups — design, control and instrumentation, protection, telecommunications etc. — are in close contact with the operating staff, and are responsible for ensuring that problems encountered on any part of the Board's system are also dealt with at other locations vulnerable to similar difficulties. Since the pre-construction phase of Torness, a formal system has been in operation to ensure that all modifications at Hunterston B, however small, are taken into account by engineers engaged on Torness design and construction, both at SSEB and NNC. It is a time-consuming and tortuous process, but it is only by such attention to detail that the systematic improvement of standards will be achieved.

The documentation required to support an operating nuclear power station has become an industry in itself. The records required for use during the lifetime of Torness, including manuals and backing reports etc., cover around one million pages. This indicates the range of effort involved in establishing and updating project documentation and gives strong support for the carefully controlled pursuit of a programme of plant replication.

12.6 Consolidation

The SSEB have made steady progress throughout the past 25 years in developing their resources for control of power-station construction, culminating in the successful progress at Torness. This has been part of a national electricity supply industry effort, also exemplified by the outstanding performance of CEGB on Drax completion as well as on the current Heysham 2 project, and by NSHEB at Peterhead. It is well recognised that the client, with responsibility for operation and safety throughout the life of a power plant, as well as carrying the overwhelming share of financial risk, must have overall project-management responsibility. Within this principle an effective agreement has been reached with NNC for the delegation of work on the nuclear island at Torness. The construction working relationship between CEGB, SSEB and NNC throughout the joint Heysham 2/Torness design contract and the subsequent manufacturing and construction phases has been of vital importance.

This Chapter has tried to give an outline of the depth of experience and effort devoted in recent years to developing power-plant project engineering. Careful attention is being given to site-management procedures and industrial relations, and the standing of construction operatives has improved. There has been extensive rationalisation of manufacturing industry, and in the main it is now well structured and equipped, if a meaningful understanding can be established on the future UK work plan.

It would be unfortunate if the opportunity is not now taken to build on experience and existing expertise through a period of consolidation in the home market. The present turbine generators, with gross output up to 750 MW and more, offer a suitable size for many years ahead. There is a sound design base for coal-fired boilers up to this size. Like any new system, the AGR has had some difficult experiences, and performance of the first five stations to the three original and very different designs, however good, will always reflect to varying degrees their prototype nature. Current operation of Hinkley/Hunterston and design and construction experience on Heysham 2/Torness give confidence, however, that the massive effort put into 'ironing out' detailed problems has taken this latest design to maturity. Heysham 2/Torness offer a sound base for an on-going AGR nuclear programme with replication to the benefit of the industry and the electricity consumer. It would be regrettable if the effort which has gone into the evolution of this design were thrown away at the stage when it can be replicated and developed as required to realise the full potential of the system. The inherent safety features of AGR offer scope for good public acceptability at a time when, throughout the world, much thinking is going into the design of alternative reactor systems.

Consolidating on construction of existing designs of commercial stations, at least to the end of the century, would enable development to be more effectively focused on the next generation of plant for installation thereafter.

Index

Team P 131